Springer Undergraduate Mathematics Series

Advisory Board

For other titles published in this series, go to
www.springer.com/series/3423

N.H. Bingham · John M. Fry

Regression

Linear Models in Statistics

 Springer

N.H. Bingham
Imperial College, London
UK
nick.bingham@btinternet.com

John M. Fry
University of East London
UK
frymaths@googlemail.com

Springer Undergraduate Mathematics Series ISSN 1615-2085
ISBN 978-1-84882-968-8   e-ISBN 978-1-84882-969-5
DOI 10.1007/978-1-84882-969-5
Springer London Dordrecht Heidelberg New York

British Library Cataloguing in Publication Data
A catalogue record for this book is available from the British Library

Library of Congress Control Number: 2010935297

Mathematics Subject Classification (2010): 62J05, 62J10, 62J12, 97K70

Cover design: Deblik

Springer is part of Springer Science+Business Media (www.springer.com)

To James, Ruth and Tom

Nick

To my parents Ingrid Fry and Martyn Fry

John

Preface

The subject of regression, or of the linear model, is central to the subject of statistics. It concerns what can be said about some quantity of interest, which we may not be able to measure, starting from information about one or more other quantities, in which we may not be interested but which we can measure. We model our variable of interest as a linear combination of these variables (called covariates), together with some error. It turns out that this simple prescription is very flexible, very powerful and useful.

If only because regression is inherently a subject in two or more dimensions, it is not the first topic one studies in statistics. So this book should not be the first book in statistics that the student uses. That said, the statistical prerequisites we assume are modest, and will be covered by any first course on the subject: ideas of sample, population, variation and randomness; the basics of parameter estimation, hypothesis testing, p–values, confidence intervals etc.; the standard distributions and their uses (normal, Student t, Fisher F and chi-square – though we develop what we need of F and chi-square for ourselves).

Just as important as a first course in statistics is a first course in probability. Again, we need nothing beyond what is met in any first course on the subject: random variables; probability distribution and densities; standard examples of distributions; means, variances and moments; some prior exposure to moment-generating functions and/or characteristic functions is useful but not essential (we include all we need here). Our needs are well served by John Haigh's book *Probability models* in the SUMS series, Haigh (2002).

Since the terms regression and linear model are largely synonymous in statistics, it is hardly surprising that we make extensive use of linear algebra and matrix theory. Again, our needs are well served within the SUMS series, in the two books by Blyth and Robertson, *Basic linear algebra* and *Further linear algebra*, Blyth and Robertson (2002a), (2002b). We make particular use of the material developed there on sums of orthogonal projections. It will be a pleasure for those familiar with this very attractive material from pure mathematics to see it being put to good use in statistics.

Practical implementation of much of the material of this book requires computer assistance – that is, access to one of the many specialist statistical packages. Since we assume that the student has already taken a first course in statistics, for which this is also true, it is reasonable for us to assume here too that the student has some prior knowledge of and experience with a statistical package. As with any other modern student text on statistics, one is here faced with various choices. One does not want to tie the exposition too tightly to any one package; one cannot cover all packages, and shouldn't try – but one wants to include some specifics, to give the text focus. We have relied here mainly on S-Plus/R®.[1]

Most of the contents are standard undergraduate material. The boundary between higher-level undergraduate courses and Master's level courses is not a sharp one, and this is reflected in our style of treatment. We have generally included complete proofs except in the last two chapters on more advanced material: Chapter 8, on Generalised Linear Models (GLMs), and Chapter 9, on special topics. One subject going well beyond what we cover – Time Series, with its extensive use of autoregressive models – is commonly taught at both undergraduate and Master's level in the UK. We have included in the last chapter some material, on non-parametric regression, which – while no harder – is perhaps as yet more commonly taught at Master's level in the UK.

In accordance with the very sensible SUMS policy, we have included exercises at the end of each chapter (except the last), as well as worked examples. One then has to choose between making the book more student-friendly, by including solutions, or more lecturer-friendly, by not doing so. We have nailed our colours firmly to the mast here by including full solutions to all exercises. We hope that the book will nevertheless be useful to lecturers also (e.g., in inclusion of references and historical background).

Rather than numbering equations, we have labelled important equations acronymically (thus the normal equations are (NE), etc.), and included such equation labels in the index. Within proofs, we have occasionally used local numbering of equations: $(*)$, (a), (b) etc.

In pure mathematics, it is generally agreed that the two most attractive subjects, at least at student level, are complex analysis and linear algebra. In statistics, it is likewise generally agreed that the most attractive part of the subject is

---

[1] S+, S-PLUS, S+FinMetrics, S+EnvironmentalStats, S+SeqTrial, S+SpatialStats, S+Wavelets, S+ArrayAnalyzer, S-PLUS Graphlets, Graphlet, Trellis, and Trellis Graphics are either trademarks or registered trademarks of Insightful Corporation in the United States and/or other countries. Insightful Corporation1700 Westlake Avenue N, Suite 500Seattle, Washington 98109 USA.

regression and the linear model. It is also extremely useful. This lovely combination of good mathematics and practical usefulness provides a counter-example, we feel, to the opinion of one of our distinguished colleagues. Mathematical statistics, Professor x opines, combines the worst aspects of mathematics with the worst aspects of statistics. We profoundly disagree, and we hope that the reader will disagree too.

The book has been influenced by our experience of learning this material, and teaching it, at a number of universities over many years, in particular by the first author's thirty years in the University of London and by the time both authors spent at the University of Sheffield. It is a pleasure to thank Charles Goldie and John Haigh for their very careful reading of the manuscript, and Karen Borthwick and her colleagues at Springer for their kind help throughout this project. We thank our families for their support and forbearance.

NHB, JMF

Imperial College, London and the University of East London, March 2010

Contents

1
Linear Regression

1.1 Introduction

When we first meet Statistics, we encounter random quantities (random variables, in probability language, or variates, in statistical language) one at a time. This suffices for a first course. Soon however we need to handle more than one random quantity at a time. Already we have to think about how they are related to each other.

Let us take the simplest case first, of two variables. Consider first the two extreme cases.

At one extreme, the two variables may be independent (unrelated). For instance, one might result from laboratory data taken last week, the other might come from old trade statistics. The two are unrelated. Each is *uninformative* about the other. They are best looked at separately. What we have here are really *two* one-dimensional problems, rather than one two-dimensional problem, and it is best to consider matters in these terms.

At the other extreme, the two variables may be essentially the same, in that each is *completely informative* about the other. For example, in the Centigrade (Celsius) temperature scale, the freezing point of water is 0° and the boiling point is 100°, while in the Fahrenheit scale, freezing point is 32° and boiling point is 212° (these bizarre choices are a result of Fahrenheit choosing as his origin of temperature the lowest temperature he could achieve in the laboratory, and recognising that the body is so sensitive to temperature that a hundredth of the freezing-boiling range as a unit is inconveniently large for everyday,

N.H. Bingham and J.M. Fry, *Regression: Linear Models in Statistics*,
Springer Undergraduate Mathematics Series, DOI 10.1007/978-1-84882-969-5_1,
© Springer-Verlag London Limited 2010

non-scientific use, unless one resorts to decimals). The transformation formulae are accordingly

$$C = (F - 32) \times 5/9, \qquad F = C \times 9/5 + 32.$$

While both scales remain in use, this is purely for convenience. To look at temperature in both Centigrade and Fahrenheit together for scientific purposes would be silly. Each is *completely informative* about the other. A plot of one against the other would lie *exactly* on a straight line. While apparently a two–dimensional problem, this would really be only *one* one-dimensional problem, and so best considered as such.

We are left with the typical and important case: two–dimensional data, $(x_1, y_1), \ldots, (x_n, y_n)$ say, where each of the x and y variables is *partially but not completely informative about the other.*

Usually, our interest is on one variable, y say, and we are interested in what knowledge of the other – x – tells us about y. We then call y the *response variable*, and x the *explanatory variable*. We know more about y knowing x than not knowing x; thus knowledge of x explains, or accounts for, part but not all of the variability we see in y. Another name for x is the *predictor* variable: we may wish to use x to predict y (the prediction will be an uncertain one, to be sure, but better than nothing: there is information content in x about y, and we want to use this information). A third name for x is the *regressor*, or regressor variable; we will turn to the reason for this name below. It accounts for why the whole subject is called *regression*.

The first thing to do with any data set is to look at it. We subject it to exploratory data analysis (EDA); in particular, we plot the graph of the n data points (x_i, y_i). We can do this by hand, or by using a statistical package: Minitab®,[1] for instance, using the command `Regression`, or S-Plus/R® by using the command `lm` (for linear model – see below).

Suppose that what we observe is a scatter plot that seems roughly linear. That is, there seems to be a systematic component, which is linear (or roughly so – linear to a first approximation, say) and an error component, which we think of as perturbing this in a random or unpredictable way. Our job is to fit a line through the data – that is, to estimate the systematic linear component.

For illustration, we recall the first case in which most of us meet such a task – experimental verification of Ohm's Law (G. S. Ohm (1787-1854), in 1826). When electric current is passed through a conducting wire, the current (in amps) is proportional to the applied potential difference or voltage (in volts), the constant of proportionality being the inverse of the *resistance* of the wire

---

[1] Minitab®, Quality Companion by Minitab®, Quality Trainer by Minitab®, Quality. Analysis. Results® and the Minitab logo are all registered trademarks of Minitab, Inc., in the United States and other countries.

(in ohms). One measures the current observed for a variety of voltages (the more the better). One then attempts to fit a line through the data, observing with dismay that, because of experimental error, no three of the data points are exactly collinear. A typical schoolboy solution is to use a perspex ruler and fit by eye. Clearly a more systematic procedure is needed. We note in passing that, as no current flows when no voltage is applied, one may restrict to lines through the origin (that is, lines with zero intercept) – by no means the typical case.

1.2 The Method of Least Squares

The required general method – the Method of Least Squares – arose in a rather different context. We know from Newton's *Principia* (Sir Isaac Newton (1642–1727), in 1687) that planets, the Earth included, go round the sun in elliptical orbits, with the Sun at one focus of the ellipse. By cartesian geometry, we may represent the ellipse by an algebraic equation of the second degree. This equation, though quadratic in the variables, is *linear* in the coefficients. How many coefficients p we need depends on the choice of coordinate system – in the range from two to six. We may make as many astronomical observations of the planet whose orbit is to be determined as we wish – the more the better, n say, where n is large – much larger than p. This makes the system of equations for the coefficients grossly over-determined, *except* that all the observations are polluted by experimental error. We need to tap the information content of the large number n of readings to make the best estimate we can of the small number p of parameters.

Write the equation of the ellipse as

$$a_1 x_1 + a_2 x_2 + \ldots = 0.$$

Here the a_j are the *coefficients*, to be found or estimated, and the x_j are those of x^2, xy, y^2, x, y, 1 that we need in the equation of the ellipse (we will always need 1, unless the ellipse degenerates to a point, which is not the case here). For the ith point, the left-hand side above will be 0 if the fit is exact, but ϵ_i say (denoting the ith error) in view of the observational errors. We wish to keep the errors ϵ_i small; we wish also to put positive and negative ϵ_i on the same footing, which we may do by looking at the squared errors ϵ_i^2. A measure of the discrepancy of the fit is the sum of these squared errors, $\sum_{i=1}^{n} \epsilon_i^2$. The Method of Least Squares is to choose the coefficients a_j so as to minimise this sums of squares,

$$SS := \sum_{i=1}^{n} \epsilon_i^2.$$

As we shall see below, this may readily and conveniently be accomplished.

The Method of Least Squares was discovered independently by two workers, both motivated by the above problem of fitting planetary orbits. It was first published by Legendre (A. M. Legendre (1752–1833), in 1805). It had also been discovered by Gauss (C. F. Gauss (1777–1855), in 1795); when Gauss published his work in 1809, it precipitated a priority dispute with Legendre.

Let us see how to implement the method. We do this first in the simplest case, the fitting of a straight line

$$y = a + bx$$

by least squares through a data set $(x_1, y_1), \ldots, (x_n, y_n)$. Accordingly, we choose a, b so as to minimise the *sum of squares*

$$SS := \sum_{i=1}^{n} \epsilon_i^2 = \sum_{i=1}^{n} (y_i - a - bx_i)^2.$$

Taking $\partial SS/\partial a = 0$ and $\partial SS/\partial b = 0$ gives

$$\partial SS/\partial a := -2\sum_{i=1}^{n} e_i = -2\sum_{i=1}^{n} (y_i - a - bx_i),$$
$$\partial SS/\partial b := -2\sum_{i=1}^{n} x_i e_i = -2\sum_{i=1}^{n} x_i(y_i - a - bx_i).$$

To find the minimum, we equate both these to zero:

$$\sum_{i=1}^{n} (y_i - a - bx_i) = 0 \quad \text{and} \quad \sum_{i=1}^{n} x_i(y_i - a - bx_i) = 0.$$

This gives two simultaneous linear equations in the two unknowns a, b, called the *normal equations*. Using the 'bar' notation

$$\bar{x} := \frac{1}{n}\sum_{i=1}^{n} x_i.$$

Dividing both sides by n and rearranging, the normal equations are

$$a + b\bar{x} = \bar{y} \quad \text{and} \quad a\bar{x} + b\overline{x^2} = \overline{xy}.$$

Multiply the first by $\bar{x}$ and subtract from the second:

$$b = \frac{\overline{xy} - \bar{x}.\bar{y}}{\overline{x^2} - (\bar{x})^2},$$

and then

$$a = \bar{y} - b\bar{x}.$$

We will use this bar notation systematically. We call $\bar{x} := \frac{1}{n}\sum_{i=1}^{n} x_i$ the *sample mean*, or average, of $x_1, \ldots, x_n$, and similarly for $\bar{y}$. In this book (though not all others!), the *sample variance* is defined as the average, $\frac{1}{n}\sum_{i=1}^{n}(x_i - \bar{x})^2$, of $(x_i - \bar{x})^2$, written s_x^2 or s_{xx}. Then using linearity of average, or 'bar',

$$s_x^2 = s_{xx} = \overline{(x - \bar{x})^2} = \overline{x^2 - 2x.\bar{x} + \bar{x}^2} = \overline{(x^2)} - 2\bar{x}.\bar{x} + (\bar{x})^2 = \overline{(x^2)} - (\bar{x})^2,$$

since $\overline{x}.\overline{x} = (\overline{x})^2$. Similarly, the *sample covariance* of x and y is defined as the average of $(x - \overline{x})(y - \overline{y})$, written s_{xy}. So

$$
\begin{aligned}
s_{xy} &= \overline{(x - \overline{x})(y - \overline{y})} = \overline{xy - x.\overline{y} - \overline{x}.y + \overline{x}.\overline{y}} \\
&= \overline{(xy)} - \overline{x}.\overline{y} - \overline{x}.\overline{y} + \overline{x}.\overline{y} = \overline{(xy)} - \overline{x}.\overline{y}.
\end{aligned}
$$

Thus the slope b is given by the *sample correlation coefficient*

$$
b = s_{xy}/s_{xx},
$$

the ratio of the sample covariance to the sample x-variance. Using the alternative 'sum of squares' notation

$$
S_{xx} := \sum_{i=1}^{n} (x_i - \overline{x})^2, \qquad S_{xy} := \sum_{i=1}^{n} (x_i - \overline{x})(y_i - \overline{y}),
$$

$$
b = S_{xy}/S_{xx}, \qquad a = \overline{y} - b\overline{x}.
$$

The line – the *least-squares line* that we have fitted – is $y = a + bx$ with this a and b, or

$$
y - \overline{y} = b(x - \overline{x}), \qquad b = s_{xy}/s_{xx} = S_{xy}/S_{xx}. \qquad (SRL)
$$

It is called the *sample regression line*, for reasons which will emerge later.

Notice that the line goes through the point $(\overline{x}, \overline{y})$ – the *centroid*, or centre of mass, of the scatter diagram $(x_1, y_1), \ldots, (x_n, y_n)$.

Note 1.1

We will see later that if we assume that the errors are *independent* and identically distributed (which we abbreviate to iid) and normal, $N(0, \sigma^2)$ say, then these formulas for a and b also give the maximum likelihood estimates. Further, $100(1 - \alpha)\%$ confidence intervals in this case can be calculated from points $\hat{a}$ and $\hat{b}$ as

$$
\begin{aligned}
a &= \hat{a} \pm t_{n-2}(1 - \alpha/2)s\sqrt{\frac{\sum x_i^2}{nS_{xx}}}, \\
b &= \hat{b} \pm \frac{t_{n-2}(1 - \alpha/2)s}{\sqrt{S_{xx}}},
\end{aligned}
$$

where $t_{n-2}(1 - \alpha/2)$ denotes the $1 - \alpha/2$ quantile of the Student t distribution with $n - 2$ degrees of freedom and s is given by

$$
s = \sqrt{\frac{1}{n-2}\left(S_{yy} - \frac{S_{xy}^2}{S_{xx}}\right)}.
$$

Example 1.2

We fit the line of best fit to model $y = $ Height (in inches) based on $x = $ Age (in years) for the following data:
$x=(14,\ 13,\ 13,\ 14,\ 14,\ 12,\ 12,\ 15,\ 13,\ 12,\ 11,\ 14,\ 12,\ 15,\ 16,\ 12,\ 15,\ 11,\ 15)$,
$y=(69,\ 56.5,\ 65.3,\ 62.8,\ 63.5,\ 57.3,\ 59.8,\ 62.5,\ 62.5,\ 59.0,\ 51.3,\ 64.3,\ 56.3,\ 66.5,\ 72.0,\ 64.8,\ 67.0,\ 57.5,\ 66.5)$.

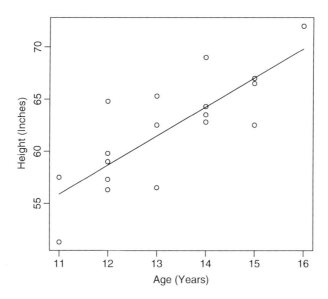

Figure 1.1   Scatter plot of the data in Example 1.2 plus fitted straight line

One may also calculate S_{xx} and S_{xy} as

$$S_{xx} = \sum x_i y_i - n\overline{x}\overline{y},$$
$$S_{xy} = \sum x_i^2 - n\overline{x}^2.$$

Since $\sum x_i y_i = 15883, \overline{x} = 13.316, \overline{y} = 62.337, \sum x_i^2 = 3409, n = 19$, we have that

$$b = \frac{15883 - 19(13.316)(62.337)}{3409 - 19(13.316^2)} = 2.787\ (3\text{ d.p.}).$$

Rearranging, we see that a becomes $62.33684 - 2.787156(13.31579) = 25.224$. This model suggests that the children are growing by just under three inches per year. A plot of the observed data and the fitted straight line is shown in Figure 1.1 and appears reasonable, although some deviation from the fitted straight line is observed.

1.2.1 Correlation version

The *sample correlation coefficient* $r = r_{xy}$ is defined as

$$r = r_{xy} := \frac{s_{xy}}{s_x s_y},$$

the quotient of the sample covariance and the product of the sample standard deviations. Thus r is dimensionless, unlike the other quantities encountered so far. One has (see Exercise 1.1)

$$-1 \leq r \leq 1,$$

with equality if and only if (iff) all the points $(x_1, y_1), \ldots, (x_n, y_n)$ lie on a straight line. Using $s_{xy} = r_{xy} s_x s_y$ and $s_{xx} = s_x^2$, we may alternatively write the sample regression line as

$$y - \overline{y} = b(x - \overline{x}), \qquad b = r_{xy} s_y / s_x. \qquad (SRL)$$

Note also that the slope b has the same sign as the sample covariance and sample correlation coefficient. These will be approximately the population covariance and correlation coefficient for large n (see below), so will have slope near zero when y and x are uncorrelated – in particular, when they are independent, and will have positive (negative) slope when x, y are positively (negatively) correlated.

We now have *five* parameters in play: two means, μ_x and μ_y, two variances σ_x^2 and σ_y^2 (or their square roots, the standard deviations σ_x and σ_y), and one correlation, ρ_{xy}. The two means are measures of *location*, and serve to identify the point – (μ_x, μ_y), or its sample counterpart, $(\overline{x}, \overline{y})$ – which serves as a natural choice of *origin*. The two variances (or standard deviations) are measures of *scale*, and serve as natural units of length along coordinate axes centred at this choice of origin. The correlation, which is dimensionless, serves as a measure of *dependence*, or *linkage*, or *association*, and indicates how closely y depends on x – that is, how informative x is about y. Note how differently these behave under affine transformations, $x \mapsto ax + b$. The mean transforms linearly:

$$E(ax + b) = aEx + b;$$

the variance transforms by

$$\text{var}(ax + b) = a^2 \text{var}(x);$$

the correlation is unchanged – it is *invariant* under affine transformations.

1.2.2 Large-sample limit

When $x_1, \ldots, x_n$ are independent copies of a random variable x, and x has mean Ex, the Law of Large Numbers says that

$$\bar{x} \to Ex \qquad (n \to \infty).$$

See e.g. Haigh (2002), §6.3. There are in fact several versions of the Law of Large Numbers (LLN). The Weak LLN (or WLLN) gives convergence in probability (for which see e.g. Haigh (2002). The Strong LLN (or SLLN) gives convergence with probability one (or 'almost surely', or 'a.s.'); see Haigh (2002) for a short proof under stronger moment assumptions (fourth moment finite), or Grimmett and Stirzaker (2001), §7.5 for a proof under the minimal condition – existence of the mean. While one should bear in mind that the SLLN holds only off some exceptional set of probability zero, we shall feel free to state the result as above, with this restriction understood. Note the content of the SLLN: thinking of a random variable as its mean plus an error, *independent errors tend to cancel when one averages.* This is essentially what makes Statistics work: the basic technique in Statistics is *averaging.*

All this applies similarly with x replaced by y, x^2, y^2, xy, when all these have means. Then

$$s_x^2 = s_{xx} = \overline{x^2} - \left(\overline{x}^2\right) \to E\left(x^2\right) - (Ex)^2 = var(x),$$

the population variance – also written $\sigma_x^2 = \sigma_{xx}$ – and

$$s_{xy} = \overline{xy} - \overline{x}.\overline{y} \to E(xy) - Ex.Ey = \text{cov}(x, y),$$

the population covariance – also written σ_{xy}. Thus as the sample size n increases, the sample regression line

$$y - \overline{y} = b(x - \overline{x}), \qquad b = s_{xy}/s_{xx}$$

tends to the line

$$y - Ey = \beta(x - Ex), \qquad \beta = \sigma_{xy}/\sigma_{xx}. \qquad (PRL)$$

This – its population counterpart – is accordingly called the *population regression line.*

Again, there is a version involving correlation, this time the *population correlation coefficient*

$$\rho = \rho_{xy} := \frac{\sigma_{xy}}{\sigma_x \sigma_y} :$$

$$y - Ey = \beta(x - Ex), \qquad \beta = \rho_{xy}\sigma_y/\sigma_x. \qquad (PRL)$$

Note 1.3

The following illustration is worth bearing in mind here. Imagine a school Physics teacher, with a class of twenty pupils; they are under time pressure revising for an exam, he is under time pressure marking. He divides the class into ten pairs, gives them an experiment to do over a double period, and withdraws to do his marking. Eighteen pupils gang up on the remaining two, the best two in the class, and threaten them into agreeing to do the experiment for them. This pair's results are then stolen by the others, who to disguise what has happened change the last two significant figures, say. Unknown to all, the best pair's instrument was dropped the previous day, and was reading way too high – so the *first* significant figures in their results, and hence all the others, were wrong. In this example, the insignificant 'rounding errors' in the last significant figures *are* independent and *do* cancel – but no significant figures are correct for any of the ten pairs, because of the strong dependence between the ten readings. Here the tenfold replication is only apparent rather than real, and is valueless. We shall see more serious examples of correlated errors in Time Series in §9.4, where high values tend to be succeeded by high values, and low values tend to be succeeded by low values.

1.3 The origins of regression

The modern era in this area was inaugurated by Sir Francis Galton (1822–1911), in his book *Hereditary genius – An enquiry into its laws and consequences* of 1869, and his paper 'Regression towards mediocrity in hereditary stature' of 1886. Galton's real interest was in intelligence, and how it is inherited. But intelligence, though vitally important and easily recognisable, is an elusive concept – human ability is infinitely variable (and certainly multi–dimensional!), and although numerical measurements of general ability exist (intelligence quotient, or IQ) and can be measured, they can serve only as a proxy for intelligence itself. Galton had a passion for measurement, and resolved to study something that *could* be easily measured; he chose human height. In a classic study, he measured the heights of 928 adults, born to 205 sets of parents. He took the average of the father's and mother's height ('mid-parental height') as the predictor variable x, and height of offspring as response variable y. (Because men are statistically taller than women, one needs to take the gender of the offspring into account. It is conceptually simpler to treat the sexes separately – and focus on sons, say – though Galton actually used an adjustment factor to compensate for women being shorter.) When he displayed his data in tabular form, Galton noticed that it showed *elliptical contours* – that is, that squares in the

(x, y)-plane containing equal numbers of points seemed to lie approximately on ellipses. The explanation for this lies in the *bivariate normal distribution*; see §1.5 below. What is most relevant here is Galton's interpretation of the sample and population regression lines (SRL) and (PRL). In (PRL), σ_x and σ_y are measures of *variability* in the parental and offspring generations. There is no reason to think that variability of height is changing (though *mean* height has visibly increased from the first author's generation to his children). So (at least to a first approximation) we may take these as equal, when (PRL) simplifies to

$$y - Ey = \rho_{xy}(x - Ex). \qquad (PRL)$$

Hence Galton's celebrated interpretation: for every inch of height above (or below) the average, the parents transmit to their children *on average* ρ inches, where ρ is the population correlation coefficient between parental height and offspring height. A further generation will introduce a further factor ρ, so the parents will transmit – again, *on average* – ρ^2 inches to their grandchildren. This will become ρ^3 inches for the great-grandchildren, and so on. Thus for every inch of height above (or below) the average, the parents transmit to their descendants after n generations *on average* ρ^n inches of height. Now

$$0 < \rho < 1$$

($\rho > 0$ as the genes for tallness or shortness are transmitted, and parental and offspring height are positively correlated; $\rho < 1$ as $\rho = 1$ would imply that parental height is *completely* informative about offspring height, which is patently not the case). So

$$\rho^n \to 0 \qquad (n \to \infty):$$

the effect of each inch of height above or below the mean is damped out with succeeding generations, and disappears in the limit. Galton summarised this as 'Regression towards mediocrity in hereditary stature', or more briefly, *regression towards the mean* (Galton originally used the term *reversion* instead, and indeed the term *mean reversion* still survives). This explains the name of the whole subject.

Note 1.4

1. We are more interested in intelligence than in height, and are more likely to take note of the corresponding conclusion for intelligence.

2. Galton found the conclusion above depressing – as may be seen from his use of the term mediocrity (to call someone average may be factual, to call

them mediocre is disparaging). Galton had a typically Victorian enthusiasm for *eugenics* – the improvement of the race. Indeed, the senior chair in Statistics in the UK (or the world), at University College London, was originally called the Galton Chair of Eugenics. This was long before the term eugenics became discredited as a result of its use by the Nazis.

3. The above assumes *random mating*. This is a reasonable assumption to make for height: height is not particularly important, while choice of mate is very important, and so few people choose their life partner with height as a prime consideration. Intelligence is quite another matter: intelligence *is* important. Furthermore, we can all observe the tendency of intelligent people to prefer and seek out each others' company, and as a natural consequence, to mate with them preferentially. This is an example of *assortative mating*. It is, of course, the best defence for intelligent people who wish to transmit their intelligence to posterity against regression to the mean. What this in fact does is to stratify the population: intelligent assortative maters are still subject to regression to the mean, but it is to a different mean – not the general population mean, but the mean among the social group in question – graduates, the learned professions or whatever.

1.4 Applications of regression

Before turning to the underlying theory, we pause to mention a variety of contexts in which regression is of great practical use, to illustrate why the subject is worth study in some detail.

1. *Examination scores.*

   This example may be of particular interest to undergraduates! The context is that of an elite institution of higher education. The proof of elite status is an excess of well-qualified applicants. These have to be ranked in merit order in some way. Procedures differ in detail, but in broad outline all relevant pieces of information – A Level scores, UCAS forms, performance in interview, admissions officer's assessment of potential etc. – are used, coded in numerical form and then combined according to some formula to give a numerical score. This is used as the predictor variable x, which measures the quality of *incoming students*; candidates are ranked by score, and places filled on merit, top down, until the quota is reached. At the end of the course, students graduate, with a classified degree. The task of the Examiners' Meeting is to award classes of degree. While at the margin this involves detailed discussion of individual cases, it is usual to table among the papers for the meeting a numerical score for each candidate, obtained by combining the relevant pieces of information – performance on the examinations taken throughout the course, assessed course-work etc. – into a numerical score, again according to some formula. This score is y, the response variable, which measures the quality of *graduating students*. The question is how well the institution picks students – that is, how good a *predictor* of eventual performance y the incoming score x is. Of course, the most important single factor here is the innate ability and personality of the individual student, plus the quality of their school education. These will be powerfully influential on *both x and y*. But they are not directly measurable, while x is, so x serves here as a proxy for them. These underlying factors remain unchanged during the student's study, and are the most important determinant of y. However, other factors intervene. Some students come to university if anything under-prepared, grow up and find their feet, and get steadily better. By contrast, some students arrive if anything over-prepared (usually as a result of expensively purchased 'cramming') and revert to their natural level of performance, while some others arrive studious and succumb to the temptations of wine, women (or men) and song, etc. The upshot is that, while x serves as a good proxy for the ability and intelligence which really matter, there is a considerable amount of unpredictability, or *noise*, here.

The question of how well institutions pick students is of great interest, to several kinds of people:

a) admissions tutors to elite institutions of higher education,

b) potential students and their parents,

c) the state, which largely finances higher education (note that in the UK in recent years, a monitoring body, OFFA – the Office for Fair Access, popularly referred to as Oftoff – has been set up to monitor such issues).

2. *Height.*

Although height is of limited importance, proud parents are consumed with a desire to foresee the future for their offspring. There are various rules of thumb for predicting the eventual future height as an adult of a small child (roughly speaking: measure at age two and double – the details vary according to sex). This is of limited practical importance nowadays, but we note in passing that some institutions or professions (the Brigade of Guards etc.) have upper and lower limits on heights of entrants.

3. *Athletic Performance*

    a) *Distance.*

Often an athlete competes at two different distances. These may be half-marathon and marathon (or ten miles and half-marathon) for the longer distances, ten kilometres and ten miles – or 5k and 10k – for the middle distances; for track, there are numerous possible pairs: 100m and 200m, 200m and 400m, 400m and 800m, 800m and 1500m, 1500m and 5,000m, 5,000m and 10,000m. In each case, what is needed – by the athlete, coach, commentator or follower of the sport – is an indication of how informative a time x over one distance is on time y over the other.

    b) *Age.*

An athlete's career has three broad phases. In the first, one completes growth and muscle development, and develops cardio-vascular fitness as the body reacts to the stresses of a training regime of running. In the second, the plateau stage, one attains one's best performances. In the third, the body is past its best, and deteriorates gradually with age. Within this third phase, age is actually a good predictor: the Rule of Thumb for ageing marathon runners (such as the first author) is that every extra year costs about an extra minute on one's marathon time.

4. *House Prices and Earnings.*

Under normal market conditions, the most important single predictor variable for house prices is earnings. The second most important predictor variable is interest rates: earnings affect the purchaser's ability to raise finance, by way of mortgage, interest rates affect ability to pay for it by servicing the mortgage. This example, incidentally, points towards the use of two predictor variables rather than one, to which we shall return below. (Under the abnormal market conditions that prevail following the Crash of 2008, or Credit Crunch, the two most relevant factors are availability of mortgage finance (which involves liquidity, credit, etc.), and confidence (which involves economic confidence, job security, unemployment, etc.).)

1.5 The Bivariate Normal Distribution

Recall two of the key ingredients of statistics:

(a) *The normal distribution, $N(\mu, \sigma^2)$:*

$$f(x) = \frac{1}{\sigma\sqrt{2\pi}} \exp\left\{-\frac{(x-\mu)^2}{2\sigma^2}\right\},$$

which has mean $EX = \mu$ and variance $var X = \sigma^2$.

(b) *Linear regression by the method of least squares* – above.

This is for *two-dimensional* (or bivariate) data $(X_1, Y_1), \ldots, (X_n, Y_n)$. Two questions arise:

(i) Why linear?

(ii) What (if any) is the two-dimensional analogue of the normal law?

Writing

$$\phi(x) := \frac{1}{\sqrt{2\pi}} \exp\left\{-\frac{1}{2}x^2\right\}$$

for the standard normal density, $\int$ for $\int_{-\infty}^{\infty}$, we shall need

(i) *recognising normal integrals:*

a) $\int \phi(x)dx = 1$ (*'normal density'*),

b) $\int x\phi(x)dx = 0$ (*'normal mean'* - or, *'symmetry'*),

c) $\int x^2\phi(x)dx = 1$ (*'normal variance'*),

(ii) *completing the square*: as for solving quadratic equations!

In view of the work above, we need an analogue in *two* dimensions of the normal distribution $N(\mu, \sigma^2)$ in one dimension. Just as in one dimension we need *two* parameters, μ and σ, in two dimensions we must expect to need *five*, by the above.

Consider the following bivariate density:

$$f(x, y) = c\exp\left\{-\frac{1}{2}Q(x, y)\right\},$$

where c is a constant, Q a positive definite quadratic form in x and y. Specifically:

$$c = \frac{1}{2\pi\sigma_1\sigma_2\sqrt{1-\rho^2}},$$

$$Q = \frac{1}{1-\rho^2}\left[\left(\frac{x-\mu_1}{\sigma_1}\right)^2 - 2\rho\left(\frac{x-\mu_1}{\sigma_1}\right)\left(\frac{y-\mu_2}{\sigma_2}\right) + \left(\frac{y-\mu_2}{\sigma_2}\right)^2\right].$$

Here $\sigma_i > 0$, μ_i are real, $-1 < \rho < 1$. Since f is clearly non-negative, to show that f is a (probability density) function (in two dimensions), it suffices to show that f integrates to 1:

$$\int_{-\infty}^{\infty}\int_{-\infty}^{\infty} f(x,y)\, dx\, dy = 1, \qquad \text{or} \qquad \int\int f = 1.$$

Write

$$f_1(x) := \int_{-\infty}^{\infty} f(x,y)\, dy, \qquad f_2(y) := \int_{-\infty}^{\infty} f(x,y)\, dx.$$

Then to show $\int\int f = 1$, we need to show $\int_{-\infty}^{\infty} f_1(x)\, dx = 1$ (or $\int_{-\infty}^{\infty} f_2(y)\, dy = 1$). Then f_1, f_2 are densities, in *one* dimension. If $f(x,y) = f_{X,Y}(x,y)$ is the *joint* density of *two* random variables X, Y, then $f_1(x)$ is the density $f_X(x)$ of X, $f_2(y)$ the density $f_Y(y)$ of Y (f_1, f_2, or f_X, f_Y, are called the *marginal* densities of the *joint* density f, or $f_{X,Y}$).

To perform the integrations, we have to *complete the square*. We have the algebraic identity

$$(1-\rho^2)Q \equiv \left[\left(\frac{y-\mu_2}{\sigma_2}\right) - \rho\left(\frac{x-\mu_1}{\sigma_1}\right)\right]^2 + (1-\rho^2)\left(\frac{x-\mu_1}{\sigma_1}\right)^2$$

(reducing the number of occurrences of y to 1, as we intend to integrate out y first). Then (taking the terms free of y out through the y-integral)

$$f_1(x) = \frac{\exp\left(-\frac{1}{2}(x-\mu_1)^2/\sigma_1^2\right)}{\sigma_1\sqrt{2\pi}}\int_{-\infty}^{\infty}\frac{1}{\sigma_2\sqrt{2\pi}\sqrt{1-\rho^2}}\exp\left(\frac{-\frac{1}{2}(y-c_x)^2}{\sigma_2^2(1-\rho^2)}\right)dy,$$

$$(*)$$

where

$$c_x := \mu_2 + \rho\frac{\sigma_2}{\sigma_1}(x-\mu_1).$$

The integral is 1 ('normal density'). So

$$f_1(x) = \frac{\exp\left(-\frac{1}{2}(x-\mu_1)^2/\sigma_1^2\right)}{\sigma_1\sqrt{2\pi}},$$

which integrates to 1 ('normal density'), proving

Fact 1. $f(x, y)$ is a joint density function (two-dimensional), with marginal density functions $f_1(x), f_2(y)$ (one-dimensional).

So we can write

$$f(x, y) = f_{X,Y}(x, y), \qquad f_1(x) = f_X(x), \qquad f_2(y) = f_Y(y).$$

Fact 2. X, Y are normal: X is $N(\mu_1, \sigma_1^2)$, Y is $N(\mu_2, \sigma_2^2)$. For, we showed $f_1 = f_X$ to be the $N(\mu_1, \sigma_1^2)$ density above, and similarly for Y by symmetry.

Fact 3. $EX = \mu_1, EY = \mu_2, \text{var } X = \sigma_1^2, \text{var } Y = \sigma_2^2$.

This identifies four out of the five parameters: two means μ_i, two variances σ_i^2.

Next, recall the definition of conditional probability:

$$P(A|B) := P(A \cap B)/P(B).$$

In the *discrete* case, if X, Y take possible values x_i, y_j with probabilities $f_X(x_i), f_Y(y_j)$, (X, Y) takes possible values (x_i, y_j) with corresponding probabilities $f_{X,Y}(x_i, y_j)$:

$$f_X(x_i) = P(X = x_i) = \Sigma_j P(X = x_i, Y = y_j) = \Sigma_j f_{X,Y}(x_i, y_j).$$

Then the *conditional* distribution of Y *given* $X = x_i$ is

$$f_{Y|X}(y_j|x_i) = \frac{P(Y = y_j, X = x_i)}{P(X = x_i)} = \frac{f_{X,Y}(x_i, y_j)}{\Sigma_j f_{X,Y}(x_i, y_j)},$$

and similarly with X, Y interchanged.

In the *density* case, we have to replace *sums* by *integrals*. Thus the conditional *density* of Y *given* $X = x$ is (see e.g. Haigh (2002), Def. 4.19, p. 80)

$$f_{Y|X}(y|x) := \frac{f_{X,Y}(x, y)}{f_X(x)} = \frac{f_{X,Y}(x, y)}{\int_{-\infty}^{\infty} f_{X,Y}(x, y)\, dy}.$$

Returning to the bivariate normal:

Fact 4. The conditional distribution of y given $X = x$ is

$$N\left(\mu_2 + \rho\frac{\sigma_2}{\sigma_1}(x - \mu_1), \quad \sigma_2^2\left(1 - \rho^2\right)\right).$$

Proof

Go back to completing the square (or, return to (∗) with $\int$ and dy deleted):

$$f(x, y) = \frac{\exp\left\{-\frac{1}{2}(x - \mu_1)^2/\sigma_1^2\right\}}{\sigma_1\sqrt{2\pi}} \cdot \frac{\exp\left\{-\frac{1}{2}(y - c_x)^2/\left(\sigma_2^2(1 - \rho^2)\right)\right\}}{\sigma_2\sqrt{2\pi}\sqrt{1 - \rho^2}}.$$

The first factor is $f_1(x)$, by Fact 1. So, $f_{Y|X}(y|x) = f(x,y)/f_1(x)$ is the second factor:

$$f_{Y|X}(y|x) = \frac{1}{\sqrt{2\pi}\sigma_2\sqrt{1-\rho^2}} \exp\left\{\frac{-(y-c_x)^2}{2\sigma_2^2(1-\rho^2)}\right\},$$

where c_x is the linear function of x given below $(*)$. □

This not only completes the proof of Fact 4 but gives
Fact 5. The conditional mean $E(Y|X = x)$ is *linear* in x:

$$E(Y|X = x) = \mu_2 + \rho\frac{\sigma_2}{\sigma_1}(x - \mu_1).$$

Note 1.5

1. This simplifies when X and Y are equally variable, $\sigma_1 = \sigma_2$:

$$E(Y|X = x) = \mu_2 + \rho(x - \mu_1)$$

(recall $EX = \mu_1, EY = \mu_2$). Recall that in Galton's height example, this says: for every inch of mid-parental height above/below the average, $x - \mu_1$, the parents pass on to their child, *on average*, ρ inches, and continuing in this way: *on average*, after n generations, each inch above/below average becomes *on average* ρ^n inches, and $\rho^n \to 0$ as $n \to \infty$, giving *regression towards the mean*.

2. This line is the population regression line (PRL), the population version of the sample regression line (SRL).

3. The relationship in Fact 5 can be generalised (§4.5): a population regression function – more briefly, a regression – is a *conditional mean*.
   This also gives

Fact 6. The conditional variance of Y given $X = x$ is

$$var(Y|X = x) = \sigma_2^2\left(1 - \rho^2\right).$$

Recall (Fact 3) that the variability (= variance) of Y is $varY = \sigma_2^2$. By Fact 5, the variability remaining in Y when X is given (i.e., not accounted for by knowledge of X) is $\sigma_2^2(1 - \rho^2)$. Subtracting, the variability of Y which *is* accounted for by knowledge of X is $\sigma_2^2\rho^2$. That is, ρ^2 is the *proportion of the variability* of Y accounted for by knowledge of X. So ρ is a measure of the *strength of association* between Y and X.

Recall that the *covariance* is defined by

$$
\begin{aligned}
\operatorname{cov}(X, Y) \quad &:= \quad E[(X - EX)(Y - EY)] = E[(X - \mu_1)(Y - \mu_2)], \\
&= \quad E(XY) - (EX)(EY),
\end{aligned}
$$

and the *correlation coefficient* ρ, or $\rho(X, Y)$, defined by

$$
\rho = \rho(X, Y) := \frac{\operatorname{cov}(X, Y)}{\sqrt{\operatorname{var} X}\sqrt{\operatorname{var} Y}} = \frac{E[(X - \mu_1)(Y - \mu_2)]}{\sigma_1 \sigma_2}
$$

is the usual measure of the strength of association between X and Y ($-1 \le \rho \le 1$; $\rho = \pm 1$ iff one of X, Y is a function of the other). That this is consistent with the use of the symbol ρ for a parameter in the density $f(x, y)$ is shown by the fact below.

Fact 7. If $(X, Y)^T$ is bivariate normal, the correlation coefficient of X, Y is ρ.

Proof

$$
\rho(X, Y) := E\left[\left(\frac{X - \mu_1}{\sigma_1}\right)\left(\frac{Y - \mu_2}{\sigma_2}\right)\right] = \int \int \left(\frac{x - \mu_1}{\sigma_1}\right)\left(\frac{y - \mu_2}{\sigma_2}\right) f(x, y) dx dy.
$$

Substitute for $f(x, y) = c \exp(-\frac{1}{2}Q)$, and make the change of variables $u := (x - \mu_1)/\sigma_1$, $v := (y - \mu_2)/\sigma_2$:

$$
\rho(X, Y) = \frac{1}{2\pi\sqrt{1 - \rho^2}} \int \int uv \exp\left(\frac{-[u^2 - 2\rho uv + v^2]}{2(1 - \rho^2)}\right) du\, dv.
$$

Completing the square as before, $[u^2 - 2\rho uv + v^2] = (v - \rho u)^2 + (1 - \rho^2)u^2$. So

$$
\rho(X, Y) = \frac{1}{\sqrt{2\pi}} \int u \exp\left(-\frac{u^2}{2}\right) du. \frac{1}{\sqrt{2\pi}\sqrt{1 - \rho^2}} \int v \exp\left(-\frac{(v - \rho u)^2}{2(1 - \rho^2)}\right) dv.
$$

Replace v in the inner integral by $(v - \rho u) + \rho u$, and calculate the two resulting integrals separately. The first is zero ('normal mean', or symmetry), the second is ρu ('normal density'). So

$$
\rho(X, Y) = \frac{1}{\sqrt{2\pi}}.\rho \int u^2 \exp\left(-\frac{u^2}{2}\right) du = \rho
$$

('normal variance'), as required.   □

This completes the identification of all five parameters in the bivariate normal distribution: two means μ_i, two variances σ_i^2, one correlation ρ.

Note 1.6

1. The above holds for $-1 < \rho < 1$; always, $-1 \leq \rho \leq 1$, by the Cauchy-Schwarz inequality (see e.g. Garling (2007) p.15, Haigh (2002) Ex 3.20 p.86, or Howie (2001) p.22 and Exercises 1.1-1.2). In the limiting cases $\rho = \pm 1$, one of X, Y is then a linear function of the other: $Y = aX + b$, say, as in the temperature example (Fahrenheit and Centigrade). The situation is not really two-dimensional: we can (and should) use only *one* of X and Y, reducing to a one-dimensional problem.

2. The slope of the regression line $y = c_x$ is $\rho\sigma_2/\sigma_1 = (\rho\sigma_1\sigma_2)/(\sigma_1^2)$, which can be written as $\mathrm{cov}(X,Y)/\mathrm{var}X = \sigma_{12}/\sigma_{11}$, or σ_{12}/σ_1^2: the line is

$$y - EY = \frac{\sigma_{12}}{\sigma_{11}}(x - EX).$$

This is the *population* version (what else?!) of the *sample regression line*

$$y - \bar{y} = \frac{s_{XY}}{s_{XX}}(x - \bar{x}),$$

familiar from linear regression.

The case $\rho = \pm 1$ – apparently two-dimensional, but really one-dimensional – is *singular*; the case $-1 < \rho < 1$ (genuinely two-dimensional) is *non-singular*, or (see below) *full rank*.

We note in passing

Fact 8. The bivariate normal law has *elliptical contours*.

For, the contours are $Q(x, y) = const$, which are ellipses (as Galton found).

Moment Generating Function (MGF).   Recall (see e.g. Haigh (2002), §5.2) the definition of the moment generating function (MGF) of a random variable X. This is the function

$$M(t), \quad \text{or} \quad M_X(t) := E \exp\{tX\}$$

for t real, and such that the expectation (typically a summation or integration, which may be infinite) converges (absolutely). For X normal $N(\mu, \sigma^2)$,

$$M(t) = \frac{1}{\sigma\sqrt{2\pi}} \int e^{tx} \exp\left(-\frac{1}{2}(x - \mu)^2/\sigma^2\right) dx.$$

Change variable to $u := (x - \mu)/\sigma$:

$$M(t) = \frac{1}{\sqrt{2\pi}} \int \exp\left(\mu t + \sigma u t - \frac{1}{2}u^2\right) du.$$

Completing the square,

$$M(t) = e^{\mu t} \frac{1}{\sqrt{2\pi}} \int \exp\left(-\frac{1}{2}(u - \sigma t)^2\right) \, du.e^{\frac{1}{2}\sigma^2 t^2},$$

or $M_X(t) = \exp(\mu t + \frac{1}{2}\sigma^2 t^2)$ (recognising that the central term on the right is $1 -$ 'normal density') . So $M_{X-\mu}(t) = \exp(\frac{1}{2}\sigma^2 t^2)$. Then (check)

$$\mu = EX = M_X'(0), \text{var } X = E[(X - \mu)^2] = M_{X-\mu}''(0).$$

Similarly in the bivariate case: the MGF is

$$M_{X,Y}(t_1, t_2) := E\exp(t_1 X + t_2 Y).$$

In the bivariate normal case:

$$
\begin{aligned}
M(t_1, t_2) &= E(\exp(t_1 X + t_2 Y)) = \int \int \exp(t_1 x + t_2 y) f(x, y) \, dx \, dy \\
&= \int \exp(t_1 x) f_1(x) \, dx \int \exp(t_2 y) f(y|x) \, dy.
\end{aligned}
$$

The inner integral is the MGF of $Y|X = x$, which is $N(c_x, \sigma_2^2, (1 - \rho^2))$, so is $\exp(c_x t_2 + \frac{1}{2}\sigma_2^2(1 - \rho^2)t_2^2)$. By Fact 5

$$c_x t_2 = [\mu_2 + \rho \frac{\sigma_2}{\sigma_1}(x - \mu_1)]t_2,$$

so $M(t_1, t_2)$ is equal to

$$\exp\left(t_2\mu_2 - t_2\frac{\sigma_2}{\sigma_1}\mu_1 + \frac{1}{2}\sigma_2^2\left(1 - \rho^2\right)t_2^2\right) \int \exp\left(\left[t_1 + t_2\rho\frac{\sigma_2}{\sigma_1}\right]x\right) f_1(x) \, dx.$$

Since $f_1(x)$ is $N(\mu_1, \sigma_1^2)$, the inner integral is a normal MGF, which is thus

$$\exp(\mu_1[t_1 + t_2\rho\frac{\sigma_2}{\sigma_1}] + \frac{1}{2}\sigma_1^2[\ldots]^2).$$

Combining the two terms and simplifying, we obtain

Fact 9. The joint MGF is

$$M_{X,Y}(t_1, t_2) = M(t_1, t_2) = \exp\left(\mu_1 t_1 + \mu_2 t_2 + \frac{1}{2}\left[\sigma_1^2 t_1^2 + 2\rho\sigma_1\sigma_2 t_1 t_2 + \sigma_2^2 t_2^2\right]\right).$$

Fact 10. X, Y are independent iff $\rho = 0$.

Proof

For densities: X, Y are independent iff the joint density $f_{X,Y}(x, y)$ *factorises* as the *product* of the marginal densities $f_X(x).f_Y(y)$ (see e.g. Haigh (2002), Cor. 4.17).

For MGFs: X, Y are independent iff the joint MGF $M_{X,Y}(t_1, t_2)$ *factorises* as the *product* of the marginal MGFs $M_X(t_1).M_Y(t_2)$. From Fact 9, this occurs iff $\rho = 0$. $\qquad\square$

Note 1.7

1. X, Y independent implies X, Y uncorrelated ($\rho = 0$) in general (when the correlation exists). The converse is *false* in general, but *true*, by Fact 10, in the bivariate normal case.

2. *Characteristic functions (CFs).* The *characteristic function*, or CF, of X is

$$\phi_X(t) := E(e^{itX}).$$

Compared to the MGF, this has the drawback of involving complex numbers, but the great advantage of always existing for t real. Indeed,

$$\left|\phi_X(t)\right| = \left|E(e^{itX})\right| \leq E\left|(e^{itX})\right| = E1 = 1.$$

By contrast, the expectation defining the MGF $M_X(t)$ may diverge for some real t (as we shall see in §2.1 with the chi-square distribution.) For background on CFs, see e.g. Grimmett and Stirzaker (2001) §5.7. For our purposes one may pass from MGF to CF by formally replacing t by it (though one actually needs analytic continuation – see e.g. Copson (1935), §4.6 – or Cauchy's Theorem – see e.g. Copson (1935), §6.7, or Howie (2003), Example 9.19). Thus for the univariate normal distribution $N(\mu, \sigma^2)$ the CF is

$$\phi_X(t) = \exp\left\{i\mu t - \frac{1}{2}\sigma^2 t^2\right\}$$

and for the bivariate normal distribution the CF of X, Y is

$$\phi_{X,Y}(t_1, t_2) = \exp\left\{i\mu_1 t_1 + i\mu_2 t_2 - \frac{1}{2}\left[\sigma_1^2 t_1^2 + 2\rho\sigma_1\sigma_2 t_1 t_2 + \sigma_2 t_2^2\right]\right\}.$$

1.6 Maximum Likelihood and Least Squares

By Fact 4, the conditional distribution of y given $X = x$ is

$$N(\mu_2 + \rho\frac{\sigma_2}{\sigma_1}(x - \mu_1), \quad \sigma_2^2(1 - \rho^2)).$$

Thus y is decomposed into two components, a *linear trend* in x – the systematic part – and a normal error, with mean zero and constant variance – the random part. Changing the notation, we can write this as

$$y = a + bx + \epsilon, \qquad \epsilon \sim N(0, \sigma^2).$$

With n values of the predictor variable x, we can similarly write

$$y_i = a + bx_i + \epsilon_i, \qquad \epsilon_i \sim N(0, \sigma^2).$$

To complete the specification of the model, we need to specify the dependence or correlation structure of the errors $\epsilon_1, \ldots, \epsilon_n$. This can be done in various ways (see Chapter 4 for more on this). Here we restrict attention to the simplest and most important case, where the errors ϵ_i are iid:

$$y_i = a + bx_i + \epsilon_i, \qquad \epsilon_i \quad \text{iid} \quad N(0, \sigma^2). \tag{$*$}$$

This is the basic model for simple linear regression.

Since each y_i is now normally distributed, we can write down its density. Since the y_i are *independent*, the joint density of $y_1, \ldots, y_n$ *factorises* as the product of the marginal (separate) densities. This joint density, regarded as a function of the *parameters*, a, b and σ, is called the *likelihood*, L (one of many contributions by the great English statistician R. A. Fisher (1890-1962), later Sir Ronald Fisher, in 1912). Thus

$$\begin{aligned} L &= \frac{1}{\sigma^n (2\pi)^{\frac{1}{2}n}} \prod_{i=1}^{n} \exp\{-\frac{1}{2}(y_i - a - bx_i)^2/\sigma^2\} \\ &= \frac{1}{\sigma^n (2\pi)^{\frac{1}{2}n}} \exp\{-\frac{1}{2}\sum_{i=1}^{n}(y_i - a - bx_i)^2/\sigma^2\}. \end{aligned}$$

Fisher suggested choosing as our estimates of the parameters the values that maximise the likelihood. This is the Method of Maximum Likelihood; the resulting estimators are the maximum likelihood estimators or MLEs. Now maximising the likelihood L and maximising its logarithm $\ell := \log L$ are the same, since the function log is increasing. Since

$$\ell := \log L = -\frac{1}{2}n \log 2\pi - n \log \sigma - \frac{1}{2}\sum_{i=1}^{n}(y_i - a - bx_i)^2/\sigma^2,$$

so far as maximising with respect to a and b are concerned (leaving σ to one side for the moment), this is the same as minimising the sum of squares $SS := \sum_{i=1}^{n}(y_i - a - bx_i)^2$ – just as in the Method of Least Squares. Summarising:

Theorem 1.8

For the normal model $(*)$, the Method of Least Squares and the Method of Maximum Likelihood are equivalent ways of estimating the parameters a and b.

It is interesting to note here that the Method of Least Squares of Legendre and Gauss belongs to the early nineteenth century, whereas Fisher's Method of Maximum Likelihood belongs to the early twentieth century. For background on the history of statistics in that period, and an explanation of the 'long pause' between least squares and maximum likelihood, see Stigler (1986).

There remains the estimation of the parameter σ, equivalently the variance σ^2. Using maximum likelihood as above gives

$$\partial \ell / \partial \sigma = \frac{-n}{\sigma} + \frac{1}{\sigma^3} \sum\nolimits_{i=1}^{n} (y_i - a - bx_i)^2 = 0,$$

or

$$\sigma^2 = \frac{1}{n} \sum\nolimits_{i=1}^{n} (y_i - a - bx_i)^2.$$

At the maximum, a and b have their maximising values $\hat{a}$, $\hat{b}$ as above, and then the maximising value $\hat{\sigma}$ is given by

$$\hat{\sigma}^2 = \frac{1}{n} \sum\nolimits_{1}^{n} (y_i - \hat{a} - \hat{b}x_i)^2 = \frac{1}{n} \sum\nolimits_{1}^{n} (y_i - \hat{y}_i)^2.$$

Note that the sum of squares SS above involves unknown parameters, a and b. Because these are unknown, one cannot calculate this sum of squares numerically from the data. In the next section, we will meet other sums of squares, which can be calculated from the data – that is, which are functions of the data, or *statistics*. Rather than proliferate notation, we will again denote the largest of these sums of squares by SS; we will then break this down into a sum of smaller sums of squares (giving a *sum of squares decomposition*). In Chapters 3 and 4, we will meet multidimensional analogues of all this, which we will handle by matrix algebra. It turns out that all sums of squares will be expressible as quadratic forms in normal variates (since the parameters, while unknown, are constant, the distribution theory of sums of squares with and without unknown parameters is the same).

1.7 Sums of Squares

Recall the sample regression line in the form

$$y = \bar{y} + b(x - \bar{x}), \qquad b = s_{xy}/s_{xx} = S_{xy}/S_{xx}. \qquad (SRL)$$

We now ask how much of the variation in y is accounted for by knowledge of x – or, as one says, by regression. The data are y_i. The fitted values are $\hat{y}_i$, the left-hand sides above with x on the right replaced by x_i. Write

$$y_i - \bar{y} = (y_i - \hat{y}_i) + (\hat{y}_i - \bar{y}),$$

square both sides and add. On the left, we get

$$SS := \sum_{i=1}^{n} (y_i - \overline{y})^2,$$

the *total sum of squares* or *sum of squares* for short. On the right, we get three terms:

$$SSR := \sum_i (\hat{y}_i - \overline{y})^2,$$

which we call the *sum of squares for regression*,

$$SSE := \sum_i (y_i - \hat{y}_i)^2,$$

the *sum of squares for error* (since this sum of squares measures the errors between the fitted values on the regression line and the data), and a cross term

$$\sum_i (y_i - \hat{y}_i)(\hat{y}_i - \overline{y}) = n\frac{1}{n}\sum_i (y_i - \hat{y}_i)(\hat{y}_i - \overline{y}) = n.\overline{(y - \hat{y})(y - \overline{y})}.$$

By (SRL), $\hat{y}_i - \overline{y} = b(x_i - \overline{x})$ with $b = S_{xy}/S_{xx} = S_{xy}/S_x^2$, and

$$y_i - \hat{y} = (y_i - \overline{y}) - b(x_i - \overline{x}).$$

So the right above is n times

$$\frac{1}{n}\sum_i b(x_i - \overline{x})[(y_i - \overline{y}) - b(x_i - \overline{x})] = bS_{xy} - b^2 S_x^2 = b\left(S_{xy} - bS_x^2\right) = 0,$$

as $b = S_{xy}/S_x^2$. Combining, we have

Theorem 1.9

$$SS = SSR + SSE.$$

In terms of the sample correlation coefficient r^2, this yields as a corollary

Theorem 1.10

$$r^2 = SSR/SS, \qquad 1 - r^2 = SSE/SS.$$

Proof

It suffices to prove the first.

$$\frac{SSR}{SS} = \frac{\sum(\hat{y}_i - \overline{y})^2}{\sum(y_i - \overline{y})^2} = \frac{\sum b^2(x_i - \overline{x})^2}{\sum(y_i - \overline{y})^2} = \frac{b^2 S_x^2}{S_y^2} = \frac{S_{xy}^2}{S_x^4} \cdot \frac{S_x^2}{S_y^2} = \frac{S_{xy}^2}{S_x^2 S_y^2} = r^2,$$

as $b = S_{xy}/S_x^2$.   □

The interpretation is that $r^2 = SSR/SS$ is the proportion of variability in y accounted for by knowledge of x, that is, by regression (and $1 - r^2 = SSE/SS$ is that unaccounted for by knowledge of x, that is, by error). This is just the sample version of what we encountered in §1.5 on the bivariate normal distribution, where (see below Fact 6 in §1.5) ρ^2 has the interpretation of the proportion of variability in y accounted for by knowledge of x. Recall that r^2 tends to ρ^2 in the large-sample limit, by the Law of Large Numbers, so the population theory of §1.5 is the large-sample limit of the sample theory here.

Example 1.11

We wish to predict y, winning speeds (mph) in a car race, given the year x, by a linear regression. The data for years one to ten are y=(140.3, 143.1, 147.4, 151.4, 144.3, 151.2, 152.9, 156.9, 155.7, 157.7). The estimates for a and b now become $\hat{a} = 139.967$ and $\hat{b} = 1.841$. Assuming normally distributed errors in our regression model means that we can now calculate confidence intervals for the parameters and express a level of uncertainty around these estimates. In this case the formulae for 95% confidence intervals give (135.928, 144.005) for a and (1.190, 2.491) for b.

Distribution theory. Consider first the case $b = 0$, when the slope is zero, there is no linear trend, and the y_i are identically distributed, $N(a, \sigma^2)$. Then $\bar{y}$ and $y_i - \bar{y}$ are also normally distributed, with zero mean. It is perhaps surprising, but true, that $\sum (y_i - \bar{y})^2$ and $\bar{y}$ are independent; we prove this in §2.5 below. The distribution of the quadratic form $\sum (y_i - \bar{y})^2$ involves the *chi-square distribution*; see §2.1 below. In this case, SSR and SSE are independent chi-square variates, and $SS = SSR + SSE$ is an instance of *chi-square decompositions*, which we meet in §3.5.

In the general case with the slope b non-zero, there is a linear trend, and a sloping regression line is more successful in explaining the data than a flat one. One quantifies this by using a ratio of sums of squares (ratio of independent chi-squares) that *increases* when the slope b is non-zero, so large values are evidence *against* zero slope. This statistic is an *F-statistic* (§2.3: F for Fisher). Such F-tests may be used to test a large variety of such *linear hypotheses* (Chapter 6).

When b is non-zero, the $y_i - \bar{y}$ are normally distributed as before, but with *non-zero* mean. Their sum of squares $\sum (y_i - \bar{y})^2$ then has a *non-central chi-square distribution*. The theory of such distributions is omitted here, but can be found in, e.g., Kendall and Stuart (1979), Ch. 24.

1.8 Two regressors

Suppose now that we have *two* regressor variables, u and v say, for the response variable y. Several possible settings have been prefigured in the discussion above:

1. *Height.*

   Galton measured the father's height u and the mother's height v in each case, before averaging to form the mid-parental height $x := (u+v)/2$. What happens if we use u and v in place of x?

2. *Predicting grain yields.*

   Here y is the grain yield after the summer harvest. Because the *price* that the grain will fetch is determined by the balance of supply and demand, and demand is fairly inflexible while supply is unpredictable, being determined largely by the weather, it is of great economic and financial importance to be able to *predict* grain yields in advance. The two most important predictors are the amount of rainfall (in cm, u say) and sunshine (in hours, v say) during the spring growing season. Given this information at the end of spring, how can we use it to best predict yield in the summer harvest? Of course, the actual harvest is still subject to events in the future, most notably the possibility of torrential rain in the harvest season flattening the crops. Note that for the sizeable market in grain futures, such predictions are highly price-sensitive information.

3. *House prices.*

   In the example above, house prices y depended on earnings u and interest rates v. We would expect to be able to get better predictions using both these as predictors than using either on its own.

4. *Athletics times.*

   We saw that both age and distance can be used separately; one ought to be able to do better by using them together.

5. *Timber.*

   The economic value of a tree grown for timber depends on the volume of usable timber when the tree has been felled and taken to the sawmill. When choosing which trees to fell, it is important to be able to estimate this volume without needing to fell the tree. The usual predictor variables here are girth (in cm, say – measured by running a tape-measure round the trunk at some standard height – one metre, say – above the ground) and height (measured by use of a surveyor's instrument and trigonometry).

With two regressors u and v and response variable y, given a sample of size n of points $(u_1, v_1, y_1), \ldots, (u_n, v_n, y_n)$ we have to fit a least-squares *plane* – that is, we have to choose parameters a, b, c to minimise the sum of squares

$$SS := \sum_{i=1}^{n} (y_i - c - au_i - bv_i)^2.$$

Taking $\partial SS/\partial c = 0$ gives

$$\sum_{i=1}^{n} (y_i - c - au_i - bv_i) = 0 : \qquad c = \overline{y} - a\overline{u} - b\overline{v}.$$

We rewrite SS as

$$SS = \sum_{i=1}^{n} [(y_i - \overline{y}) - a(u_i - \overline{u}) - b(v_i - \overline{v})]^2.$$

Then $\partial SS/\partial a = 0$ and $\partial SS/\partial b = 0$ give

$$\sum_{i=1}^{n} (u_i - \overline{u})[(y_i - \overline{y}) - a(u_i - \overline{u}) - b(v_i - \overline{v})] = 0,$$

$$\sum_{i=1}^{n} (v_i - \overline{v})[(y_i - \overline{y}) - a(u_i - \overline{u}) - b(v_i - \overline{v})] = 0.$$

Multiply out, divide by n to turn the sums into averages, and re-arrange using our earlier notation of sample variances and sample covariance: the above equations become

$$as_{uu} + bs_{uv} = s_{yu},$$

$$as_{uv} + bs_{vv} = s_{yv}.$$

These are the *normal equations* for a and b. The determinant is

$$s_{uu}s_{vv} - s_{uv}^2 = s_{uu}s_{vv}(1 - r_{uv}^2)$$

(since $r_{uv} := s_{uv}/(s_u s_v)$). This is non-zero iff $r_{uv} \neq \pm 1$ – that is, iff the points $(u_1, v_1), \ldots, (u_n, v_n)$ are not collinear – and this is the condition for the normal equations to have a unique solution.

The extension to three or more regressors may be handled in just the same way: with p regressors we obtain p normal equations. The general case is best handled by the matrix methods of Chapter 3.

Note 1.12

As with the linear regression case, under the assumption of iid $N(0, \sigma^2)$ errors these formulas for a and b also give the maximum likelihood estimates. Further,

$100(1 - \alpha)\%$ confidence intervals can be returned routinely using standard software packages, and in this case can be calculated as

$$c = \hat{c} \pm t_{n-3}(1 - \alpha/2)s\sqrt{\frac{\sum u_i^2 \sum v_i^2 - (\sum u_i v_i)^2}{n\sum u_i^2 S_{vv} + n\sum u_i v_i [2n\overline{uv} - \sum u_i v_i] - n^2\overline{u}^2 \sum v_i^2}},$$

$$a = \hat{a} \pm t_{n-3}(1 - \alpha/2)s\sqrt{\frac{S_{vv}}{\sum u_i^2 S_{vv} + \sum u_i v_i [2n\overline{uv} - \sum u_i v_i] - n\overline{u}^2 \sum v_i^2}},$$

$$b = \hat{b} \pm t_{n-3}(1 - \alpha/2)s\sqrt{\frac{S_{uu}}{\sum u_i^2 S_{vv} + \sum u_i v_i [2n\overline{uv} - \sum u_i v_i] - n\overline{u}^2 \sum v_i^2}},$$

where

$$s = \sqrt{\frac{1}{n-3}\left(S_{yy} - \hat{a}S_{uy} - \hat{b}S_{vy}\right)};$$

see Exercise 3.10.

Note 1.13 (Joint confidence regions)

In the above, we restrict ourselves to confidence intervals for individual parameters, as is done in e.g. S-Plus/R®. One can give confidence regions for two or more parameters together, we refer for detail to Draper and Smith (1998), Ch. 5.

EXERCISES

1.1. By considering the quadratic

$$Q(\lambda) := \frac{1}{n}\sum_{i=1}^{n}(\lambda(x_i - \overline{x}) + (y_i - \overline{y}))^2,$$

show that the sample correlation coefficient r satisfies
(i) $-1 \leq r \leq 1$;
(ii) $r = \pm 1$ iff there is a linear relationship between x_i and y_i,

$$ax_i + by_i = c \quad (i = 1, \ldots, n).$$

1.2. By considering the quadratic

$$Q(\lambda) := E[(\lambda(x - \overline{x}) + (y - \overline{y}))^2],$$

show that the population correlation coefficient ρ satisfies
(i) $-1 \leq \rho \leq 1$;

(ii) $\rho = \pm 1$ iff there is a linear relationship between x and y, $ax + by = c$ with probability 1.

(These results are both instances of the Cauchy–Schwarz inequality for sums and integrals respectively.)

1.3. *The effect of ageing on athletic performance.* The data in Table 1.1 gives the first author's times for the marathon and half-marathon (in minutes).

(i) Fit the model $\log(time) = a + b\log(age)$ and give estimates and

Age	Half-marathon	Age	Marathon
46	85.62	46.5	166.87
48	84.90	47.0	173.25
49	87.88	47.5	175.17
50	87.88	49.5	178.97
51	87.57	50.5	176.63
57	90.25	54.5	175.03
59	88.40	56.0	180.32
60	89.45	58.5	183.02
61	96.38	59.5	192.33
62	94.62	60.0	191.73

Table 1.1 Data for Exercise 1.3

95% confidence intervals for a and b.

(ii) Compare your results with the runners' Rule of Thumb that, for ageing athletes, every year of age adds roughly half a minute to the half-marathon time and a full minute to the marathon time.

1.4. Look at the data for Example 1.11 on car speeds. Plot the data along with the fitted regression line. Fit the model $y = a + bx + cx^2$ and test for the significance of a quadratic term. Predict the speeds for x=(-3, 13) and compare with the actual observations of 135.9 and 158.6 respectively. Which model seems to predict best out of sample? Do your results change much when you add these two observations to your sample?

1.5. Give the solution to the normal equations for the regression model with two regressors in §1.8

1.6. Consider the data in Table 1.2 giving the first author's half-marathon times:

Age (x)	Time (y)	Age (x)	Time (y)
42	92.00	51	87.57
43	92.00	57	90.25
44	91.25	59	88.40
46	85.62	60	89.45
48	84.90	61	96.38
49	87.88	62	94.62
50	87.88	63	91.23

Table 1.2 Data for Exercise 1.6

(i) Fit the models $y = a + bx$ and $y = a + bx + cx^2$. Does the extra quadratic term appear necessary?

(ii) *Effect of club membership upon performance.* Use the following proxy $v = (0, 0, 0, 1, 1, 1, 1, 1, 1, 1, 1, 1, 1, 1)$ to gauge the effect of club membership. ($v = 1$ corresponds to being a member of a club). Consider the model $y = a + bx + cv$. How does membership of a club appear to affect athletic performance?

1.7. The following data, $y = (9.8, 11.0, 13.2, 15.1, 16.0)$ give the price index y in years one to five.

(i) Which of the models $y = a + bt$, $y = Ae^{bt}$ fits the data best?

(ii) Does the quadratic model, $y = a + bt + ct^2$ offer a meaningful improvement over the simple linear regression model?

1.8. The following data in Table 1.3 give the US population in millions. Fit a suitable model and interpret your findings.

Year	Population	Year	Population
1790	3.93	1890	62.90
1800	5.31	1900	76.00
1810	7.24	1910	92.00
1820	9.64	1920	105.70
1830	12.90	1930	122.80
1840	17.10	1940	131.70
1850	23.20	1950	151.30
1860	31.40	1960	179.30
1870	39.80	1970	203.20
1880	50.20		

Table 1.3 Data for Exercise 1.8.

1.9. *One-dimensional change-of-variable formula.* Let X be a continuous random variable with density $f_X(x)$. Let $Y = g(X)$ for some monotonic function $g(\cdot)$.

(i) Show that

$$f_Y(x) = f_X\left(g^{-1}(x)\right)\left|\frac{dg^{-1}(x)}{dx}\right|.$$

(ii) Suppose $X \sim N(\mu, \sigma^2)$. Show that $Y = e^X$ has probability density function

$$f_Y(x) = \frac{1}{\sqrt{2\pi}\sigma}\exp\left\{-\frac{(\log x - \mu)^2}{2\sigma^2}\right\}.$$

[Note that this gives the log-normal distribution, important in the Black–Scholes model of mathematical finance.]

1.10. The following exercise motivates a discussion of Student's t distribution as a normal variance mixture (see Exercise 1.11). Let $U \sim \chi_r^2$ be a chi-squared distribution with r degrees of freedom (for which see §2.1), with density

$$f_U(x) = \frac{x^{\frac{1}{2}r-1}e^{-\frac{1}{2}x}}{2^{\frac{1}{2}r}\Gamma(\frac{r}{2})}.$$

(i) Show, using Exercise 1.9 or differentiation under the integral sign that $Y = r/U$ has density

$$f_Y(x) = \frac{r^{\frac{1}{2}r}x^{-1-\frac{1}{2}r}e^{-\frac{1}{2}rx^{-1}}}{2^{\frac{1}{2}r}\Gamma(\frac{r}{2})}.$$

(ii) Show that if $X \sim \Gamma(a, b)$ with density

$$f_X(x) = \frac{x^{a-1}b^ae^{-bx}}{\Gamma(a)},$$

then $Y = X^{-1}$ has density

$$f_Y(x) = \frac{b^ax^{-1-a}e^{-b/x}}{\Gamma(a)}.$$

Deduce the value of

$$\int_0^\infty x^{-1-a}e^{-b/x}dx.$$

1.11. *Student's t distribution.* A Student t distribution $t(r)$ with r degrees of freedom can be constructed as follows:
1. Generate u from $f_Y(\cdot)$.
2. Generate x from $N(0, u)$,
where $f_Y(\cdot)$ is the probability density in Exercise 1.10 (ii). Show that

$$f_{t(r)}(x) = \frac{\Gamma\left(\frac{r}{2} + \frac{1}{2}\right)}{\sqrt{\pi r}\,\Gamma(\frac{r}{2})}\left(1 + \frac{x^2}{r}\right)^{-\frac{1}{2}(r+1)}.$$

The Student t distribution often arises in connection with the chi-square distribution (see Chapter 2). If $X \sim N(0, 1)$ and $Y \sim \chi_r^2$ with X and Y independent then

$$\frac{X}{\sqrt{Y/r}} \sim t(r).$$

2

The Analysis of Variance (ANOVA)

While the linear regression of Chapter 1 goes back to the nineteenth century, the Analysis of Variance of this chapter dates from the twentieth century, in applied work by Fisher motivated by agricultural problems (see §2.6). We begin this chapter with some necessary preliminaries, on the special distributions of Statistics needed for small-sample theory: the chi-square distributions $\chi^2(n)$ (§2.1), the Fisher F-distributions $F(m,n)$ (§2.3), and the independence of normal sample means and sample variances (§2.5). We shall generalise linear regression to multiple regression in Chapters 3 and 4 – which use the Analysis of Variance of this chapter – and unify regression and Analysis of Variance in Chapter 5 on Analysis of Covariance.

2.1 The Chi-Square Distribution

We now define the *chi-square distribution* with *n degrees of freedom* (df), $\chi^2(n)$. This is the distribution of

$$X_1^2 + \ldots + X_n^2,$$

with the X_i iid $N(0,1)$.

Recall (§1.5, Fact 9) the definition of the MGF, and also the definition of the *Gamma function*,

$$\Gamma(t) := \int_0^\infty e^{-x} x^{t-1} dx \qquad (t > 0)$$

N.H. Bingham and J.M. Fry, *Regression: Linear Models in Statistics*,
Springer Undergraduate Mathematics Series, DOI 10.1007/978-1-84882-969-5_2,
© Springer-Verlag London Limited 2010

(the integral converges for $t > 0$). One may check (by integration by parts) that

$$\Gamma(n+1) = n! \qquad (n = 0, 1, 2, \ldots),$$

so the Gamma function provides a continuous extension to the factorial. It is also needed in Statistics, as it comes into the normalisation constants of the standard distributions of small-sample theory, as we see below.

Theorem 2.1

The chi-square distribution $\chi^2(n)$ with n degrees of freedom has
(i) mean n and variance $2n$,
(ii) MGF $M(t) = 1/(1-2t)^{\frac{1}{2}n}$ for $t < \frac{1}{2}$,
(iii) density

$$f(x) = \frac{1}{2^{\frac{1}{2}n}\Gamma\left(\frac{1}{2}n\right)} . x^{\frac{1}{2}n-1} \exp\left(-\frac{1}{2}x\right) \qquad (x > 0).$$

Proof

(i) For $n = 1$, the mean is 1, because a $\chi^2(1)$ is the square of a standard normal, and a standard normal has mean 0 and variance 1. The variance is 2, because the fourth moment of a standard normal X is 3, and

$$\mathrm{var}\left(X^2\right) = E\left[\left(X^2\right)^2\right] - \left[E\left(X^2\right)\right]^2 = 3 - 1 = 2.$$

For general n, the mean is n because means add, and the variance is $2n$ because variances add over independent summands (Haigh (2002), Th 5.5, Cor 5.6).
(ii) For X standard normal, the MGF of its square X^2 is

$$M(t) := \int e^{tx^2}\phi(x)\,dx = \frac{1}{\sqrt{2\pi}}\int_{-\infty}^{\infty} e^{tx^2}e^{-\frac{1}{2}x^2}\,dx = \frac{1}{\sqrt{2\pi}}\int_{-\infty}^{\infty} e^{-\frac{1}{2}(1-2t)x^2}\,dx.$$

So the integral converges only for $t < \frac{1}{2}$; putting $y := \sqrt{1-2t}.x$ gives

$$M(t) = 1/\sqrt{1-2t} \qquad \left(t < \frac{1}{2}\right) \qquad \text{for } X \sim N(0,1).$$

Now when X, Y are independent, the MGF of their sum is the product of their MGFs (see e.g. Haigh (2002), p.103). For e^{tX}, e^{tY} are independent, and the mean of an independent product is the product of the means. Combining these, the MGF of a $\chi^2(n)$ is given by

$$M(t) = 1/(1-2t)^{\frac{1}{2}n} \qquad \left(t < \frac{1}{2}\right) \qquad \text{for } X \sim \chi^2(n).$$

(iii) First, $f(.)$ is a density, as it is non-negative, and integrates to 1:

$$
\begin{aligned}
\int f(x)\, dx &= \frac{1}{2^{\frac{1}{2}n}\Gamma\left(\frac{1}{2}n\right)} \int_0^\infty x^{\frac{1}{2}n-1} \exp\left(-\frac{1}{2}x\right)\, dx \\
&= \frac{1}{\Gamma\left(\frac{1}{2}n\right)} \int_0^\infty u^{\frac{1}{2}n-1} \exp(-u)\, du \qquad \left(u := \frac{1}{2}x\right) \\
&= 1,
\end{aligned}
$$

by definition of the Gamma function. Its MGF is

$$
\begin{aligned}
M(t) &= \frac{1}{2^{\frac{1}{2}n}\Gamma\left(\frac{1}{2}n\right)} \int_0^\infty e^{tx} x^{\frac{1}{2}n-1} \exp\left(-\frac{1}{2}x\right)\, dx \\
&= \frac{1}{2^{\frac{1}{2}n}\Gamma\left(\frac{1}{2}n\right)} \int_0^\infty x^{\frac{1}{2}n-1} \exp\left(-\frac{1}{2}x(1-2t)\right)\, dx.
\end{aligned}
$$

Substitute $u := x(1-2t)$ in the integral. One obtains

$$
M(t) = (1-2t)^{-\frac{1}{2}n} \frac{1}{2^{\frac{1}{2}n}\Gamma\left(\frac{1}{2}n\right)} \int_0^\infty u^{\frac{1}{2}n-1} e^{-u}\, du = (1-2t)^{-\frac{1}{2}n},
$$

by definition of the Gamma function. $\qquad\square$

Chi-square Addition Property. If X_1, X_2 are independent, $\chi^2(n_1)$ and $\chi^2(n_2)$, $X_1 + X_2$ is $\chi^2(n_1 + n_2)$.

Proof

$X_1 = U_1^2 + \ldots + U_{n_1}^2$, $X_2 = U_{n_1+1}^2 + \ldots + U_{n_1+n_2}^2$, with U_i iid $N(0,1)$.
So $X_1 + X_2 = U_1^2 + \cdots + U_{n_1+n_2}^2$, so $X_1 + X_2$ is $\chi^2(n_1 + n_2)$. $\qquad\square$

Chi-Square Subtraction Property. If $X = X_1 + X_2$, with X_1 and X_2 independent, and $X \sim \chi^2(n_1 + n_2)$, $X_1 \sim \chi^2(n_1)$, then $X_2 \sim \chi^2(n_2)$.

Proof

As X is the independent sum of X_1 and X_2, its MGF is the product of their MGFs. But X, X_1 have MGFs $(1-2t)^{-\frac{1}{2}(n_1+n_2)}$, $(1-2t)^{-\frac{1}{2}n_1}$. Dividing, X_2 has MGF $(1-2t)^{-\frac{1}{2}n_2}$. So $X_2 \sim \chi^2(n_2)$. $\qquad\square$

2.2 Change of variable formula and Jacobians

Recall from calculus of several variables the change of variable formula for multiple integrals. If in

$$I := \int \cdots \int_A f(x_1, \ldots, x_n) \, dx_1 \ldots dx_n = \int_A f(\mathbf{x}) \, d\mathbf{x}$$

we make a one-to-one change of variables from $\mathbf{x}$ to $\mathbf{y}$ — $\mathbf{x} = \mathbf{x}(\mathbf{y})$ or $x_i = x_i(y_1, \ldots, y_n)$ $(i = 1, \ldots, n)$ — let B be the region in $\mathbf{y}$-space corresponding to the region A in $\mathbf{x}$-space. Then

$$I = \int_A f(\mathbf{x}) \, d\mathbf{x} = \int_B f(\mathbf{x}(\mathbf{y})) \left| \frac{\partial \mathbf{x}}{\partial \mathbf{y}} \right| \, d\mathbf{y} = \int_B f(\mathbf{x}(\mathbf{y})) |J| \, d\mathbf{y},$$

where J, the determinant of partial derivatives

$$J := \frac{\partial \mathbf{x}}{\partial \mathbf{y}} = \frac{\partial(x_1, \cdots, x_n)}{\partial(y_1, \cdots, y_n)} := \det \left(\frac{\partial x_i}{\partial y_j} \right)$$

is the *Jacobian* of the transformation (after the great German mathematician C. G. J. Jacobi (1804–1851) in 1841 – see e.g. Dineen (2001), Ch. 14). Note that in one dimension, this just reduces to the usual rule for change of variables: $dx = (dx/dy).dy$. Also, if J is the Jacobian of the change of variables $\mathbf{x} \to \mathbf{y}$ above, the Jacobian $\partial \mathbf{y}/\partial \mathbf{x}$ of the inverse transformation $\mathbf{y} \to \mathbf{x}$ is J^{-1} (from the product theorem for determinants: $det(AB) = detA.detB$ – see e.g. Blyth and Robertson (2002a), Th. 8.7).

Suppose now that $\mathbf{X}$ is a random n-vector with density $f(\mathbf{x})$, and we wish to change from $\mathbf{X}$ to $\mathbf{Y}$, where $\mathbf{Y}$ corresponds to $\mathbf{X}$ as $\mathbf{y}$ above corresponds to $\mathbf{x}$: $\mathbf{y} = \mathbf{y}(\mathbf{x})$ iff $\mathbf{x} = \mathbf{x}(\mathbf{y})$. If $\mathbf{Y}$ has density $g(\mathbf{y})$, then by above,

$$P(\mathbf{X} \in A) = \int_A f(\mathbf{x}) \, d\mathbf{x} = \int_B f(\mathbf{x}(\mathbf{y})) \left| \frac{\partial \mathbf{x}}{\partial \mathbf{y}} \right| \, d\mathbf{y},$$

and also

$$P(\mathbf{X} \in A) = P(\mathbf{Y} \in B) = \int_B g(\mathbf{y}) d\mathbf{y}.$$

Since these hold for all B, the integrands must be equal, giving

$$g(\mathbf{y}) = f(\mathbf{x}(\mathbf{y})) |\partial \mathbf{x}/\partial \mathbf{y}|$$

as the density g of $\mathbf{Y}$.

In particular, if the change of variables is linear:

$$\mathbf{y} = \mathbf{A}\mathbf{x} + \mathbf{b}, \quad \mathbf{x} = \mathbf{A}^{-1}\mathbf{y} - \mathbf{A}^{-1}\mathbf{b}, \quad \partial \mathbf{y}/\partial \mathbf{x} = |\mathbf{A}|, \quad \partial \mathbf{x}/\partial \mathbf{y} = |\mathbf{A}^{-1}| = |\mathbf{A}|^{-1}.$$

2.3 The Fisher F-distribution

Suppose we have two independent random variables U and V, chi-square distributed with degrees of freedom (df) m and n respectively. We divide each by its df, obtaining U/m and V/n. The distribution of the *ratio*

$$F := \frac{U/m}{V/n}$$

will be important below. It is called the *F-distribution* with *degrees of freedom* (m, n), $F(m, n)$. It is also known as the (Fisher) *variance-ratio distribution*.

Before introducing its density, we define the *Beta function*,

$$B(\alpha, \beta) := \int_0^1 x^{\alpha-1}(1-x)^{\beta-1}dx,$$

wherever the integral converges ($\alpha > 0$ for convergence at 0, $\beta > 0$ for convergence at 1). By *Euler's integral for the Beta function*,

$$B(\alpha, \beta) = \frac{\Gamma(\alpha)\Gamma(\beta)}{\Gamma(\alpha + \beta)}$$

(see e.g. Copson (1935), §9.3). One may then show that the density of $F(m, n)$ is

$$f(x) = \frac{m^{\frac{1}{2}m}n^{\frac{1}{2}n}}{B(\frac{1}{2}m, \frac{1}{2}m)} \cdot \frac{x^{\frac{1}{2}(m-2)}}{(mx + n)^{\frac{1}{2}(m+n)}} \qquad (m, n > 0, \quad x > 0)$$

(see e.g. Kendall and Stuart (1977), §16.15, §11.10; the original form given by Fisher is slightly different).

There are two important features of this density. The first is that (to within a normalisation constant, which, like many of those in Statistics, involves ratios of Gamma functions) it behaves near zero like the power $x^{\frac{1}{2}(m-2)}$ and near infinity like the power $x^{-\frac{1}{2}n}$, and is smooth and unimodal (has one peak). The second is that, like all the common and useful distributions in Statistics, its percentage points are *tabulated*. Of course, using tables of the F-distribution involves the complicating feature that one has *two* degrees of freedom (rather than one as with the chi-square or Student t-distributions), and that these must be taken in the correct *order*. It is sensible at this point for the reader to take some time to gain familiarity with use of tables of the F-distribution, using whichever standard set of statistical tables are to hand. Alternatively, all standard statistical packages will provide percentage points of F, t, χ^2, etc. on demand. Again, it is sensible to take the time to gain familiarity with the statistical package of your choice, including use of the online Help facility.

One can derive the density of the F distribution from those of the χ^2 distributions above. One needs the formula for the density of a quotient of random variables. The derivation is left as an exercise; see Exercise 2.1. For an introduction to calculations involving the F distribution see Exercise 2.2.

2.4 Orthogonality

Recall that a square, non-singular $(n \times n)$ matrix A is *orthogonal* if its inverse is its transpose:

$$A^{-1} = A^T.$$

We now show that the property of being independent $N(0, \sigma^2)$ is preserved under an orthogonal transformation.

Theorem 2.2 (Orthogonality Theorem)

If $X = (X_1, \ldots, X_n)^T$ is an n-vector whose components are independent random variables, normally distributed with mean 0 and variance σ^2, and we change variables from X to Y by

$$Y := AX$$

where the matrix A is orthogonal, then the components Y_i of Y are again independent, normally distributed with mean 0 and variance σ^2.

Proof

We use the Jacobian formula. If $A = (a_{ij})$, since $\partial Y_i / \partial X_j = a_{ij}$, the Jacobian $\partial Y / \partial X = |A|$. Since A is orthogonal, $AA^T = AA^{-1} = I$. Taking determinants, $|A|.|A^T| = |A|.|A| = 1$: $|A| = 1$, and similarly $|A^T| = 1$. Since length is preserved under an orthogonal transformation,

$$\sum_1^n Y_i^2 = \sum_1^n X_i^2.$$

The joint density of $(X_1, \ldots, X_n)$ is, by independence, the product of the marginal densities, namely

$$f(x_1, \ldots, x_n) = \prod_{i=1}^n \frac{1}{\sqrt{2\pi}} \exp\left\{ -\frac{1}{2} x_i^2 \right\} = \frac{1}{(2\pi)^{\frac{1}{2}n}} \exp\left\{ -\frac{1}{2} \sum_1^n x_i^2 \right\}.$$

From this and the Jacobian formula, we obtain the joint density of $(Y_1, \ldots, Y_n)$ as

$$f(y_1, \ldots, y_n) = \frac{1}{(2\pi)^{\frac{1}{2}n}} \exp\left\{ -\frac{1}{2} \sum_1^n y_i^2 \right\} = \prod_1^n \frac{1}{\sqrt{2\pi}} \exp\left\{ -\frac{1}{2} y_i^2 \right\}.$$

But this is the joint density of n independent standard normals – and so $(Y_1, \ldots, Y_n)$ are independent standard normal, as claimed. $\square$

Helmert's Transformation.

There exists an orthogonal $n \times n$ matrix P with first row

$$\frac{1}{\sqrt{n}}(1, \ldots, 1)$$

(there are many such! Robert Helmert (1843–1917) made use of one when he introduced the χ^2 distribution in 1876 – see Kendall and Stuart (1977), Example 11.1 – and it is convenient to use his name here for any of them.) For, take this vector, which spans a one-dimensional subspace; take $n-1$ unit vectors not in this subspace and use the Gram–Schmidt orthogonalisation process (see e.g. Blyth and Robertson (2002b), Th. 1.4) to obtain a set of n orthonormal vectors.

2.5 Normal sample mean and sample variance

For $X_1, \ldots, X_n$ independent and identically distributed (iid) random variables, with mean μ and variance σ^2, write

$$\overline{X} := \frac{1}{n}\sum_1^n X_i$$

for the *sample mean* and

$$S^2 := \frac{1}{n}\sum_1^n (X_i - \overline{X})^2$$

for the *sample variance*.

Note 2.3

Many authors use $1/(n-1)$ rather than $1/n$ in the definition of the sample variance. This gives S^2 as an *unbiased* estimator of the population variance σ^2. But our definition emphasizes the parallel between the bar, or average, for sample quantities and the expectation for the corresponding population quantities:

$$\overline{X} = \frac{1}{n}\sum_1^n X_i \leftrightarrow EX,$$

$$S^2 = \overline{(X - \overline{X})^2} \leftrightarrow \sigma^2 = E\left[(X - EX)^2\right],$$

which is mathematically more convenient.

Theorem 2.4

If $X_1, \ldots, X_n$ are iid $N(\mu, \sigma^2)$,
(i) the sample mean $\overline{X}$ and the sample variance S^2 are independent,
(ii) $\overline{X}$ is $N(\mu, \sigma^2/n)$,
(iii) nS^2/σ^2 is $\chi^2(n-1)$.

Proof

(i) Put $Z_i := (X_i - \mu)/\sigma$, $Z := (Z_1, \ldots, Z_n)^T$; then the Z_i are iid $N(0,1)$,

$$\overline{Z} = (\overline{X} - \mu)/\sigma, \qquad nS^2/\sigma^2 = \sum_1^n (Z_i - \overline{Z})^2.$$

Also, since

$$\begin{aligned}
\sum_1^n (Z_i - \overline{Z})^2 &= \sum_1^n Z_i^2 - 2\overline{Z}\sum_1^n Z_i + n\overline{Z}^2 \\
&= \sum_1^n Z_i^2 - 2\overline{Z}.n\overline{Z} + n\overline{Z}^2 = \sum_1^n Z_i^2 - n\overline{Z}^2 : \\
\sum_1^n Z_i^2 &= \sum_1^n (Z_i - \overline{Z})^2 + n\overline{Z}^2.
\end{aligned}$$

The terms on the right above are quadratic forms, with matrices A, B say, so we can write

$$\sum_1^n Z_i^2 = Z^T A Z + Z^T B X. \qquad (*)$$

Put $W := PZ$ with P a Helmert transformation – P orthogonal with first row $(1, \ldots, 1)/\sqrt{n}$:

$$W_1 = \frac{1}{\sqrt{n}}\sum_1^n Z_i = \sqrt{n}\overline{Z}; \qquad W_1^2 = n\overline{Z}^2 = Z^T B Z.$$

So

$$\sum_2^n W_i^2 = \sum_1^n W_i^2 - W_1^2 = \sum_1^n Z_i^2 - Z^T B Z = Z^T A Z = \sum_1^n (Z_i - \overline{Z})^2 = nS^2/\sigma^2.$$

But the W_i are independent (by the orthogonality of P), so W_1 is independent of $W_2, \ldots, W_n$. So W_1^2 is independent of $\sum_2^n W_i^2$. So nS^2/σ^2 is independent of $n(\overline{X} - \mu)^2/\sigma^2$, so S^2 is independent of $\overline{X}$, as claimed.
(ii) We have $\overline{X} = (X_1 + \ldots + X_n)/n$ with X_i independent, $N(\mu, \sigma^2)$, so with MGF $\exp(\mu t + \frac{1}{2}\sigma^2 t^2)$. So X_i/n has MGF $\exp(\mu t/n + \frac{1}{2}\sigma^2 t^2/n^2)$, and $\overline{X}$ has MGF

$$\prod_1^n \exp\left(\mu t/n + \frac{1}{2}\sigma^2 t^2/n^2\right) = \exp\left(\mu t + \frac{1}{2}\sigma^2 t^2/n\right).$$

So $\overline{X}$ is $N(\mu, \sigma^2/n)$.
(iii) In $(*)$, we have on the left $\sum_1^n Z_i^2$, which is the sum of the squares of n standard normals Z_i, so is $\chi^2(n)$ with MGF $(1 - 2t)^{-\frac{1}{2}n}$. On the right, we have

two independent terms. As $\overline{Z}$ is $N(0, 1/n)$, $\sqrt{n}\overline{Z}$ is $N(0,1)$, so $n\overline{Z}^2 = Z^T BZ$ is $\chi^2(1)$, with MGF $(1-2t)^{-\frac{1}{2}}$. Dividing (as in chi-square subtraction above), $Z^T AZ = \sum_1^n (Z_i - \overline{Z})^2$ has MGF $(1-2t)^{-\frac{1}{2}(n-1)}$. So $Z^T AZ = \sum_1^n (Z_i - \overline{Z})^2$ is $\chi^2(n-1)$. So nS^2/σ^2 is $\chi^2(n-1)$. $\qquad\square$

Note 2.5

1. This is a remarkable result. We quote (without proof) that this property actually *characterises* the normal distribution: if the sample mean and sample variance are independent, then the population distribution is normal (Geary's Theorem: R. C. Geary (1896–1983) in 1936; see e.g. Kendall and Stuart (1977), Examples 11.9 and 12.7).

2. The fact that when we form the sample mean, the mean is unchanged, while the variance decreases by a factor of the sample size n, is true generally. The point of (ii) above is that normality is preserved. This holds more generally: it will emerge in Chapter 4 that normality is preserved under any linear operation.

Theorem 2.6 (Fisher's Lemma)

Let $X_1, \ldots, X_n$ be iid $N(0, \sigma^2)$. Let

$$Y_i = \sum_{j=1}^n c_{ij} X_j \qquad (i = 1, \ldots, p, \quad p < n),$$

where the row-vectors $(c_{i1}, \ldots, c_{in})$ are orthogonal for $i = 1, \ldots, p$. If

$$S^2 = \sum_1^n X_i^2 - \sum_1^p Y_i^2,$$

then
(i) S^2 is independent of $Y_1, \ldots, Y_p$,
(ii) S^2 is $\chi^2(n-p)$.

Proof

Extend the $p \times n$ matrix (c_{ij}) to an $n \times n$ orthogonal matrix $C = (c_{ij})$ by Gram–Schmidt orthogonalisation. Then put

$$Y := CX,$$

so defining $Y_1, \ldots, Y_p$ (again) and $Y_{p+1}, \ldots, Y_n$. As C is orthogonal, $Y_1, \ldots, Y_n$ are iid $N(0, \sigma^2)$, and $\sum_1^n Y_i^2 = \sum_1^n X_i^2$. So

$$S^2 = \left(\sum_1^n - \sum_1^p\right) Y_i^2 = \sum_{p+1}^n Y_i^2$$

is independent of $Y_1, \ldots, Y_p$, and S^2/σ^2 is $\chi^2(n-p)$. $\qquad\square$

2.6 One-Way Analysis of Variance

To compare two normal means, we use the Student t-test, familiar from your first course in Statistics. What about comparing r means for $r > 2$?

Analysis of Variance goes back to early work by Fisher in 1918 on mathematical genetics and was further developed by him at Rothamsted Experimental Station in Harpenden, Hertfordshire in the 1920s. The convenient acronym ANOVA was coined much later, by the American statistician John W. Tukey (1915–2000), the pioneer of exploratory data analysis (EDA) in Statistics (Tukey (1977)), and coiner of the terms hardware, software and bit from computer science.

Fisher's motivation (which arose directly from the agricultural field trials carried out at Rothamsted) was to compare yields of several varieties of crop, say – or (the version we will follow below) of one crop under several fertiliser *treatments*. He realised that if there was more variability between groups (of yields with different treatments) than within groups (of yields with the same treatment) than one would expect if the treatments were the same, then this would be evidence against believing that they were the same. In other words, Fisher set out to *compare means by analysing variability* ('variance' – the term is due to Fisher – is simply a short form of 'variability').

We write μ_i for the mean yield of the ith variety, for $i = 1, \ldots, r$. For each i, we draw n_i independent readings X_{ij}. The X_{ij} are independent, and we assume that they are normal, all with the same unknown variance σ^2:

$$X_{ij} \sim N(\mu_i, \sigma^2) \qquad (j = 1, \ldots, n_i, \quad i = 1, \ldots, r).$$

We write

$$n := \sum_1^r n_i$$

for the total sample size.

With two suffices i and j in play, we use a bullet to indicate that the suffix in that position has been averaged out. Thus we write

$$X_{i\bullet}, \quad \text{or} \quad \overline{X}_{i}, := \frac{1}{n_i} \sum_{j=1}^{n_i} X_{ij} \qquad (i = 1, \ldots, r)$$

for the ith *group mean* (the sample mean of the ith sample)

$$X_{\bullet\bullet}, \quad \text{or} \quad \overline{X}, := \frac{1}{n} \sum_{i=1}^r \sum_{j=1}^{n_i} X_{ij} = \frac{1}{n} \sum_{i=1}^r n_i X_{i\bullet}$$

for the *grand mean* and,

$$S_i^2 := \frac{1}{n_i}\sum_{j=1}^{n_i}(X_{ij} - X_{i\bullet})^2$$

for the *i*th sample variance.

Define the *total sum of squares*

$$SS := \sum_{i=1}^{r}\sum_{j=1}^{n_i}(X_{ij} - X_{\bullet\bullet})^2 = \sum_i\sum_j[(X_{ij} - X_{i\bullet}) + (X_{i\bullet} - X_{\bullet\bullet})]^2.$$

As

$$\sum_j(X_{ij} - X_{i\bullet}) = 0$$

(from the definition of $X_{i\bullet}$ as the average of the X_{ij} over j), if we expand the square above, the cross terms vanish, giving

$$
\begin{aligned}
SS &= \sum_i\sum_j(X_{ij} - X_{i\bullet})^2 \\
&+ \sum_i\sum_j(X_{ij} - X_{i\bullet})(X_{i\bullet} - X_{\bullet\bullet}) \\
&+ \sum_i\sum_j(X_{i\bullet} - X_{\bullet\bullet})^2 \\
&= \sum_i\sum_j(X_{ij} - X_{i\bullet})^2 + \sum_i\sum_j X_{i\bullet} - X_{\bullet\bullet})^2 \\
&= \sum_i n_i S_i^2 + \sum_i n_i(X_{i\bullet} - X_{\bullet\bullet})^2.
\end{aligned}
$$

The first term on the right measures the amount of variability *within* groups. The second measures the variability *between* groups. We call them the *sum of squares for error* (or *within groups*), *SSE*, also known as the *residual sum of squares*, and the *sum of squares for treatments* (or *between groups*), respectively:

$$SS = SSE + SST,$$

where

$$SSE := \sum_i n_i S_i^2, \qquad SST := \sum_i n_i(X_{i\bullet} - X_{\bullet\bullet})^2.$$

Let H_0 be the null hypothesis of no treatment effect:

$$H_0: \qquad \mu_i = \mu \qquad (i = 1, \dots, r).$$

If H_0 is true, we have merely one large sample of size n, drawn from the distribution $N(\mu, \sigma^2)$, and so

$$SS/\sigma^2 = \frac{1}{\sigma^2}\sum_i\sum_j(X_{ij} - X_{\bullet\bullet})^2 \sim \chi^2(n-1) \qquad \text{under } H_0.$$

In particular,

$$E[SS/(n-1)] = \sigma^2 \qquad \text{under } H_0.$$

Whether or not H_0 is true,

$$n_i S_i^2/\sigma^2 = \frac{1}{\sigma^2}\sum_j (X_{ij} - X_{i\bullet})^2 \sim \chi^2(n_i - 1).$$

So by the Chi-Square Addition Property

$$SSE/\sigma^2 = \sum_i n_i S_i^2/\sigma^2 = \frac{1}{\sigma^2}\sum_i \sum_j (X_{ij} - X_{i\bullet})^2 \sim \chi^2(n - r),$$

since as $n = \sum_i n_i$,

$$\sum_{i=1}^r (n_i - 1) = n - r.$$

In particular,

$$E[SSE/(n - r)] = \sigma^2.$$

Next,

$$SST := \sum_i n_i (X_{i\bullet} - X_{\bullet\bullet})^2, \quad \text{where} \quad X_{\bullet\bullet} = \frac{1}{n}\sum_i n_i X_{i\bullet}, \quad SSE := \sum_i n_i S_i^2.$$

Now S_i^2 is independent of $X_{i\bullet}$, as these are the sample variance and sample mean from the ith sample, whose independence was proved in Theorem 2.4. Also S_i^2 is independent of $X_{j\bullet}$ for $j \neq i$, as they are formed from different independent samples. Combining, S_i^2 is independent of all the $X_{j\bullet}$, so of their (weighted) average $X_{\bullet\bullet}$, so of SST, a function of the $X_{j\bullet}$ and of $X_{\bullet\bullet}$. So $SSE = \sum_i n_i S_i^2$ is also independent of SST.

We can now use the Chi-Square Subtraction Property. We have, under H_0, the independent sum

$$SS/\sigma^2 = SSE/\sigma^2 +_{ind} SST/\sigma^2.$$

By above, the left-hand side is $\chi^2(n - 1)$, while the first term on the right is $\chi^2(n - r)$. So the second term on the right must be $\chi^2(r - 1)$. This gives:

Theorem 2.7

Under the conditions above and the null hypothesis H_0 of no difference of treatment means, we have the sum-of-squares decomposition

$$SS = SSE +_{ind} SST,$$

where $SS/\sigma^2 \sim \chi^2(n - 1)$, $SSE/\sigma^2 \sim \chi^2(n - r)$ and $SSE/\sigma^2 \sim \chi^2(r - 1)$.

When we have a sum of squares, chi-square distributed, and we divide by its degrees of freedom, we will call the resulting ratio a *mean sum of squares*, and denote it by changing the SS in the name of the sum of squares to MS. Thus the mean sum of squares is

$$MS := SS/\mathrm{df}(SS) = SS/(n-1)$$

and the mean sums of squares for treatment and for error are

$$
\begin{aligned}
MST &:= SST/\mathrm{df}(SST) = SST/(r-1),\\
MSE &:= SSE/\mathrm{df}(SSE) = SSE/(n-r).
\end{aligned}
$$

By the above,

$$SS = SST + SSE;$$

whether or not H_0 is true,

$$E[MSE] = E[SSE]/(n-r) = \sigma^2;$$

under H_0,

$$E[MS] = E[SS]/(n-1) = \sigma^2, \qquad \text{and so also} \qquad E[MST]/(r-1) = \sigma^2.$$

Form the F-statistic

$$F := MST/MSE.$$

Under H_0, this has distribution $F(r-1, n-r)$. Fisher realised that comparing the size of this F-statistic with percentage points of this F-distribution gives us a way of testing the truth or otherwise of H_0. Intuitively, if the treatments do differ, this will tend to inflate SST, hence MST, hence $F = MST/MSE$. To justify this intuition, we proceed as follows. Whether or not H_0 is true,

$$
\begin{aligned}
SST &= \sum_i n_i(X_{i\bullet} - X_{\bullet\bullet})^2 = \sum_i n_i X_{i\bullet}^2 - 2X_{\bullet\bullet}\sum_i n_i X_{i\bullet} + X_{\bullet\bullet}^2 \sum_i n_i\\
&= \sum_i n_i X_{i\bullet}^2 - nX_{\bullet\bullet}^2,
\end{aligned}
$$

since $\sum_i n_i X_{i\bullet} = nX_{\bullet\bullet}$ and $\sum_i n_i = n$. So

$$
\begin{aligned}
E[SST] &= \sum_i n_i E\left[X_{i\bullet}^2\right] - nE\left[X_{\bullet\bullet}^2\right]\\
&= \sum_i n_i \left[\mathrm{var}(X_{i\bullet}) + (EX_{i\bullet})^2\right] - n\left[\mathrm{var}(X_{\bullet\bullet}) + (EX_{\bullet\bullet})^2\right].
\end{aligned}
$$

But $\mathrm{var}(X_{i\bullet}) = \sigma^2/n_i$,

$$
\begin{aligned}
\mathrm{var}(X_{\bullet\bullet}) &= \mathrm{var}(\frac{1}{n}\sum_{i=1}^r n_i X_{i\bullet}) = \frac{1}{n^2}\sum_1^r n_i^2 \mathrm{var}(X_{i\bullet}),\\
&= \frac{1}{n^2}\sum_1^r n_i^2 \sigma^2/n_i = \sigma^2/n
\end{aligned}
$$

(as $\sum_i n_i = n$). So writing

$$\bar{\mu} := \frac{1}{n}\sum_i n_i \mu_i = EX_{\bullet\bullet} = E\frac{1}{n}\sum_i n_i X_{i\bullet},$$

$$\begin{aligned} E(SST) &= \sum_1^r n_i \left[\frac{\sigma^2}{n_i} + \mu_i^2\right] - n\left[\frac{\sigma^2}{n} + \bar{\mu}^2\right] \\ &= (r-1)\sigma^2 + \sum_i n_i \mu_i^2 - n\bar{\mu}^2 \\ &= (r-1)\sigma^2 + \sum_i n_i(\mu_i - \bar{\mu})^2 \end{aligned}$$

(as $\sum_i n_i = n$, $n\bar{\mu} = \sum_i n_i \mu_i$). This gives the inequality

$$E[SST] \geq (r-1)\sigma^2,$$

with equality iff

$$\mu_i = \bar{\mu} \quad (i = 1, \ldots, r), \qquad \text{i.e.} \qquad H_0 \text{ is true.}$$

Thus when H_0 is *false*, the mean of *SST increases*, so *larger* values of *SST*, so of *MST* and of $F = MST/MSE$, are evidence *against* H_0. It is thus appropriate to use a *one-tailed* F-test, rejecting H_0 if the value F of our F-statistic is *too big*. How big is too big depends, of course, on our chosen significance level α, and hence on the tabulated value $F_{tab} := F_\alpha(r-1, n-r)$, the upper α-point of the relevant F-distribution. We summarise:

Theorem 2.8

When the null hypothesis H_0 (that all the treatment means $\mu_1, \ldots, \mu_r$ are equal) is true, the F-statistic $F := MST/MSE = (SST/(r-1))/(SSE/(n-r))$ has the F-distribution $F(r-1, n-r)$. When the null hypothesis is false, F increases. So large values of F are evidence against H_0, and we test H_0 using a one-tailed test, rejecting at significance level α if F is too big, that is, with critical region

$$F > F_{tab} = F_\alpha(r-1, n-r).$$

Model Equations for One-Way ANOVA.

$$X_{ij} = \mu_i + \epsilon_{ij} \qquad (i = 1, \ldots, r, \quad j = 1, \ldots, r), \qquad \epsilon_{ij} \text{ iid } N(0, \sigma^2).$$

Here μ_i is the *main effect* for the ith treatment, the null hypothesis is H_0: $\mu_1 = \ldots = \mu_r = \mu$, and the unknown variance σ^2 is a nuisance parameter. The point of forming the ratio in the F-statistic is to cancel this nuisance parameter σ^2, just as in forming the ratio in the Student t-statistic in one's first course in Statistics. We will return to nuisance parameters in §5.1.1 below.

Calculations.

In any calculation involving variances, there is cancellation to be made, which is worthwhile and important numerically. This stems from the definition and 'computing formula' for the variance,

$$\sigma^2 := E\left[(X - EX)^2\right] = E\left[X^2\right] - (EX)^2$$

and its sample counterpart

$$S^2 := \overline{(X - \overline{X})^2} = \overline{X^2} - \overline{X}^2.$$

Writing T, T_i for the grand total and group totals, defined by

$$T := \sum_i \sum_j X_{ij}, \qquad T_i := \sum_j X_{ij},$$

so $X_{\bullet\bullet} = T/n$, $nX_{\bullet\bullet}^2 = T^2/n$:

$$SS = \sum_i \sum_j X_{ij}^2 - T^2/n,$$

$$SST = \sum_i T_i^2/n_i - T^2/n,$$

$$SSE = SS - SST = \sum_i \sum_j X_{ij}^2 - \sum_i T_i^2/n_i.$$

These formulae help to reduce rounding errors and are easiest to use if carrying out an Analysis of Variance by hand.

It is customary, and convenient, to display the output of an Analysis of Variance by an ANOVA table, as shown in Table 2.1. (The term 'Error' can be used in place of 'Residual' in the 'Source' column.)

Source	df	SS	Mean Square	F
Treatments	$r - 1$	SST	$MST = SST/(r-1)$	MST/MSE
Residual	$n - r$	SSE	$MSE = SSE/(n-r)$	
Total	$n - 1$	SS		

Table 2.1 One-way ANOVA table.

Example 2.9

We give an example which shows how to calculate the Analysis of Variance tables by hand. The data in Table 2.2 come from an agricultural experiment. We wish to test for different mean yields for the different fertilisers. We note that

Fertiliser	Yield
A	14.5, 12.0, 9.0, 6.5
B	13.5, 10.0, 9.0, 8.5
C	11.5, 11.0, 14.0, 10.0
D	13.0, 13.0, 13.5, 7.5
E	15.0, 12.0, 8.0, 7.0
F	12.5, 13.5, 14.0, 8.0

Table 2.2 Data for Example 2.9

we have six treatments so $6 - 1 = 5$ degrees of freedom for treatments. The total number of degrees of freedom is the number of observations minus one, hence 23. This leaves 18 degrees of freedom for the within-treatments sum of squares. The total sum of squares can be calculated routinely as $\sum(y_{ij} - \bar{y}^2) = \sum y_{ij}^2 - n\bar{y}^2$, which is often most efficiently calculated as $\sum y_{ij}^2 - (1/n)\left(\sum y_{ij}\right)^2$. This calculation gives $SS = 3119.25 - (1/24)(266.5)^2 = 159.990$. The easiest next step is to calculate SST, which means we can then obtain SSE by subtraction as above. The formula for SST is relatively simple and reads $\sum_i T_i/n_i - T^2/n$, where T_i denotes the sum of the observations corresponding to the ith treatment and $T = \sum_{ij} y_{ij}$. Here this gives $SST = (1/4)(42^2 + 41^2 + 46.5^2 + 47^2 + 42^2 + 48^2) - 1/24(266.5)^2 = 11.802$. Working through, the full ANOVA table is shown in Table 2.3.

Source	df	Sum of Squares	Mean Square	F
Between fertilisers	5	11.802	2.360	0.287
Residual	18	148.188	8.233	
Total	23	159.990		

Table 2.3 One-way ANOVA table for Example 2.9

This gives a non-significant p-value compared with $F_{3,16}(0.95) = 3.239$. R calculates the p-value to be 0.914. Alternatively, we may place bounds on the p-value by looking at statistical tables. In conclusion, we have no evidence for differences between the various types of fertiliser.

In the above example, the calculations were made more simple by having equal numbers of observations for each treatment. However, the same general procedure works when this no longer continues to be the case. For detailed worked examples with unequal sample sizes see Snedecor and Cochran (1989) §12.10.

S-Plus/R®.

We briefly describe implementation of one-way ANOVA in S-Plus/R®. For background and details, see e.g. Crawley (2002), Ch. 15. Suppose we are studying the dependence of yield on treatment, as above. [Note that this requires that we set treatment to be a *factor* variable, taking discrete rather than continuous values, which can be achieved by setting `treatment <- factor(treatment)`.] Then, using `aov` as short for 'Analysis of Variance', `<-` for the assignment operator in S-Plus (read as 'goes to' or 'becomes') and `~` as short for 'depends on' or 'is regressed on', we use

```
model <- aov (yield ~ treatment)
```

to do the analysis, and ask for the summary table by

```
summary(model)
```

A complementary `anova` command is summarised briefly in Chapter 5.2.1.

2.7 Two-Way ANOVA; No Replications

In the agricultural experiment considered above, problems may arise if the growing area is not homogeneous. The plots on which the different treatments are applied may differ in fertility – for example, if a field slopes, nutrients tend to leach out of the soil and wash downhill, so lower-lying land may give higher yields than higher-lying land. Similarly, differences may arise from differences in drainage, soil conditions, exposure to sunlight or wind, crops grown in the past, etc. If such differences are not taken into account, we will be unable to distinguish between differences in yield resulting from differences in *treatment*, our object of study, and those resulting from differences in growing conditions – *plots*, for short – which are not our primary concern. In such a case, one says that treatments are *confounded* with plots – we would have no way of separating the effect of one from that of the other.

The only way out of such difficulties is to subdivide the growing area into plots, each of which can be treated as a homogeneous growing area, and then subdivide each plot and apply different treatments to the different sub-plots or blocks. In this way we will be 'comparing like with like', and avoid the pitfalls of confounding.

When allocating treatments to blocks, we may wish to *randomise*, to avoid the possibility of inadvertently introducing a treatment-block linkage. Relevant here is the subject of *design of experiments*; see §9.3.

In the sequel, we assume for simplicity that the block sizes are the same and the number of treatments is the same for each block. The model equations will now be of the form

$$X_{ij} = \mu + \alpha_i + \beta_j + \epsilon_{ij} \qquad (i = 1, \ldots, r, \quad j = 1, \ldots, n).$$

Here μ is the *grand mean* (or *overall mean*); α_i is the ith *treatment effect* (we take $\sum_i \alpha_i = 0$, otherwise this sum can – and so should – be absorbed into μ; β_j is the jth *block effect* (similarly, we take $\sum_j \beta_j = 0$); the errors ϵ_{ij} are iid $N(0, \sigma^2)$, as before.

Recall the terms $X_{i\bullet}$ from the one-way case; their counterparts here are similarly denoted $X_{\bullet j}$. Start with the algebraic identity

$$(X_{ij} - X_{\bullet\bullet}) = (X_{ij} - X_{i\bullet} - X_{\bullet j} + X_{\bullet\bullet}) + (X_{i\bullet} - X_{\bullet\bullet}) + (X_{\bullet j} - X_{\bullet\bullet}).$$

Square and add. One can check that the cross terms cancel, leaving only the squared terms. For example, $(X_{ij} - X_{i\bullet} - X_{\bullet j} + X_{\bullet\bullet})$ averages over i to $-(X_{\bullet j} - X_{\bullet\bullet})$, and over j to $-(X_{\bullet j} - X_{\bullet\bullet})$, while each of the other terms on the right involves only one of i and j, and so is unchanged when averaged over the other. One is left with

$$\sum_{i=1}^{r}\sum_{j=1}^{n}(X_{ij} - X_{\bullet\bullet})^2 = \sum_{i=1}^{r}\sum_{j=1}^{n}(X_{ij} - X_{i\bullet} - X_{\bullet j} + X_{\bullet\bullet})^2$$
$$+ n\sum_{i=1}^{r}(X_{i\bullet} - X_{\bullet\bullet})^2$$
$$+ r\sum_{j=1}^{n}(X_{\bullet j} - X_{\bullet\bullet})^2.$$

We write this as

$$SS = SSE + SST + SSB,$$

giving the total sum of squares SS as the sum of the sum of squares for error (SSE), the sum of squares for treatments (SST) (as before) and a new term, the sum of squares for *blocks*, (SSB). The degrees of freedom are, respectively, $nr - 1$ for SS (the total sample size is nr, and we lose one df in estimating σ), $r - 1$ for treatments (as before), $n - 1$ for blocks (by analogy with treatments – or equivalently, there are n block parameters β_j, but they are subject to one constraint, $\sum_j \beta_j = 0$), and $(n-1)(r-1)$ for error (to give the correct total in the df column in the table below). Independence of the three terms on the right follows by arguments similar to those in the one-way case. We can accordingly construct a two-way ANOVA table, as in Table 2.4.

Here we have *two* F-statistics, $FT := MST/MSE$ for treatment effects and $FB := MSB/MSE$ for block effects. Accordingly, we can test *two* null hypotheses, one, $H_0(T)$, for presence of a treatment effect and one, $H_0(B)$, for presence of a block effect.

Source	df	SS	Mean Square	F
Treatments	$r-1$	SST	$MST = \frac{SST}{r-1}$	MST/MSE
Blocks	$n-1$	SSB	$MSB = \frac{SSB}{n-1}$	MSB/MSE
Residual	$(r-1)(n-1)$	SSE	$MSE = \frac{SSE}{(r-1)(n-1)}$	
Total	$rn-1$	SS		

Table 2.4 Two-way ANOVA table

Note 2.10

In educational psychology (or other behavioural sciences), 'treatments' might be different questions on a test, 'blocks' might be *individuals*. We take it for granted that individuals differ. So we need not calculate MSB nor test $H_0(B)$ (though packages such as S-Plus will do so automatically). Then $H_0(T)$ as above tests for differences between mean scores on questions in a test. (Where the questions carry equal credit, such differences are undesirable – but may well be present in practice!)

Implementation. In S-Plus, the commands above extend to

```
model <- aov(yield ~ treatment + block)

summary(model)
```

Example 2.11

We illustrate the two-way Analysis of Variance with an example. We return to the agricultural example in Example 2.9, but suppose that the data can be linked to growing areas as shown in Table 2.5. We wish to test the hypothesis that there are no differences between the various types of fertiliser. The

Fertiliser	Area 1	Area 2	Area 3	Area 4
A	14.5	12.0	9.0	6.5
B	13.5	10.0	9.0	8.5
C	11.5	11.0	14.0	10.0
D	13.0	13.0	13.5	7.5
E	15.0	12.0	8.0	7.0
F	12.5	13.5	14.0	8.0

Table 2.5 Data for Example 2.11

sum-of-squares decomposition for two-way ANOVA follows in an analogous way to the one-way case. There are relatively simple formulae for SS, SST, and SSB, meaning that SSE can easily be calculated by subtraction. In detail, these formulae are

$$SS = \sum_{ij} X_{ij}^2 - \frac{1}{nr}\left(\sum X_{ij}\right)^2,$$

$$SST = \left(X_{1\bullet}^2 + \ldots + X_{r\bullet}^2\right)/n - \frac{1}{nr}\left(\sum X_{ij}\right)^2,$$

$$SSB = \left(X_{\bullet 1}^2 + \ldots + X_{\bullet n}^2\right)/r - \frac{1}{nr}\left(\sum X_{ij}\right)^2,$$

with $SSE = SS - SST - SSB$. Returning to our example, we see that

$$SS = 3119.25 - (1/24)(266.5)^2 = 159.990,$$
$$SST = (42^2 + 41^2 + 46.5^2 + 47^2 + 42^2 + 48^2)/4 - (1/24)(266.5)^2 = 11.802,$$
$$SSB = (80^2 + 71.5^2 + 67.5^2 + 47.5^2)/6 - (1/24)(266.5)^2 = 94.865.$$

By subtraction $SSE = 159.9896 - 11.80208 - 94.86458 = 53.323$. These calculations lead us to the ANOVA table in Table 2.6. Once again we have no evidence for differences amongst the 6 types of fertiliser. The variation that does occur is mostly due to the effects of different growing areas.

Source	df	S.S.	MS	F	p
Fertilisers	5	11.802	2.360	0.664	0.656
Area	3	94.865	31.622	8.895	0.001
Residual	15	53.323	3.555		
Total	23	159.990			

Table 2.6 Two-way ANOVA table for Example 2.11

2.8 Two-Way ANOVA: Replications and Interaction

In the above, we have one reading X_{ij} for each *cell*, or combination of the ith treatment and the jth block. But we may have more. Suppose we have m *replications* – independent readings – per cell. We now need three suffices rather than two. The model equations will now be of the form

$$X_{ijk} = \mu + \alpha_i + \beta_j + \gamma_{ij} + \epsilon_{ijk} \qquad (i = 1, \ldots, r, \quad j = 1, \ldots, n, \quad k = 1, \ldots, m).$$

Here the new parameters γ_{ij} measure possible *interactions* between treatment and block effects. This allows one to study situations in which effects are *not additive*. Although we use the word interaction here as a technical term in Statistics, this is fully consistent with its use in ordinary English. We are all familiar with situations where, say, a medical treatment (e.g. a drug) may interact with some aspect of our diet (e.g. alcohol). Similarly, two drugs may interact (which is why doctors must be careful in checking what medication a patient is currently taking before issuing a new prescription). Again, different alcoholic drinks may interact (folklore wisely counsels against mixing one's drinks), etc.

Arguments similar to those above lead to the following sum-of-squares decomposition:

$$\sum_{i=1}^{r}\sum_{j=1}^{n}(X_{ijk} - X_{\bullet\bullet\bullet})^2 = \sum_i\sum_j\sum_k(X_{ijk} - X_{ij\bullet})^2$$
$$+ nm\sum_i(X_{i\bullet\bullet} - X_{\bullet\bullet\bullet})^2$$
$$+ rm\sum_j(X_{\bullet j\bullet} - X_{\bullet\bullet\bullet})^2$$
$$+ m\sum_i\sum_j(X_{ij\bullet} - X_{i\bullet\bullet} - X_{\bullet j\bullet} + X_{\bullet\bullet\bullet})^2.$$

We write this as

$$SS = SSE + SST + SSB + SSI,$$

where the new term is the sum of squares for *interactions*. The degrees of freedom are $r-1$ for treatments as before, $n-1$ for blocks as before, $(r-1)(n-1)$ for interactions (the product of the effective number of parameters for treatments and for blocks), $rnm-1$ in total (there are rnm readings), and $rn(m-1)$ for error (so that the df totals on the right and left above agree).

Implementation.   The S-Plus/R® commands now become

```
model <- aov(yield ~ treatment * block)

summary(model)
```

This notation is algebraically motivated, and easy to remember. With *additive* effects, we used a $+$. We now use a $*$, suggestive of the possibility of 'product' terms representing the interactions. We will encounter many more such situations in the next chapter, when we deal with multiple regression.

The summary table now takes the form of Table 2.7. We now have *three* F-statistics, FT and FB as before, and now FI also, which we can use to test for the presence of interactions.

Source	df	SS	Mean Square	F
Treatments	$r-1$	SST	$MST = \frac{SST}{r-1}$	MST/MSE
Blocks	$n-1$	SSB	$MSB = \frac{SSB}{n-1}$	MSB/MSE
Interaction	$(r-1)(n-1)$	SSI	$MSI = \frac{SSI}{(r-1)(n-1)}$	MSI/MSE
Residual	$rn(m-1)$	SSE	$MSE = \frac{SSE}{rn(m-1)}$	
Total	$rmn-1$	SS		

Table 2.7 Two-way ANOVA table with interactions

Example 2.12

The following example illustrates the procedure for two-way ANOVA with interactions. The data in Table 2.8 link the growth of hamsters of different coat colours when fed different diets.

	Light coat	Dark coat
Diet A	6.6, 7.2	8.3, 8.7
Diet B	6.9, 8.3	8.1, 8.5
Diet C	7.9, 9.2	9.1, 9.0

Table 2.8 Data for Example 2.12

The familiar formula for the total sum of squares gives $SS = 805.2 - (97.8^2/12) = 8.13$. In a similar manner to Example 2.11, the main effects sum-of-squares calculations give

$$SST = \sum \frac{y_{i\bullet\bullet}^2}{nm} - \frac{\left(\sum_{ijk} y_{ijk}\right)^2}{rmn},$$

$$SSB = \frac{y_{\bullet j\bullet}^2}{rm} - \frac{\left(\sum_{ijk} y_{ijk}\right)^2}{rmn},$$

and in this case give $SST = (1/4)(30.8^2 + 31.8^2 + 35.2^2) - (97.8^2/12) = 2.66$ and $SSB = (1/6)(46.1^2 + 51.7^2) - (97.8^2/12) = 2.613$. The interaction sum of squares can be calculated as a sum of squares corresponding to every cell in the table once the main effects of SST and SSB have been accounted for. The calculation is

$$SSI = \frac{1}{m}\sum y_{ij\bullet}^2 - SST - SSB - \frac{\left(\sum_{ijk} y_{ijk}\right)^2}{rmn},$$

which in this example gives $SSI = (1/2)(13.8^2 + 17^2 + 15.2^2 + 16.6^2 + 17.1^2 + 18.1^2) - 2.66 - 2.613 - (97.8^2/12) = 0.687$. As before, SSE can be calculated by subtraction, and the ANOVA table is summarised in Table 2.9. The results

Source	df	SS	MS	F	p
Diet	2	2.66	1.33	3.678	0.091
Coat	1	2.613	2.613	7.226	0.036
Diet:Coat	2	0.687	0.343	0.949	0.438
Residual	5	2.17	0.362		
Total	11	8.13			

Table 2.9 Two-way ANOVA with interactions for Example 2.12.

suggest that once we take into account the different types of coat, the effect of the different diets is seen to become only borderline significant. The diet:coat interaction term is seen to be non-significant and we might consider in a subsequent analysis the effects of deleting this term from the model.

Note 2.13 (Random effects)

The model equation for two-way ANOVA with interactions is

$$y_{ijk} = \mu + \alpha_i + \beta_j + \gamma_{ij} + \epsilon_{ijk},$$

with $\sum_i \alpha_i = \sum_j \beta_j = \sum_{ij} \gamma_{ij} = 0$. Here the α_i, β_j, γ_{ij} are constants, and the randomness is in the errors ϵ_{ijk}. Suppose, however, that the β_i were themselves random (in the examination set-up above, the suffix i might refer to the ith question, and suffix j to the jth candidate; the candidates might be chosen at random from a larger population). We would then use notation such as

$$y_{ijk} = \mu + \alpha_i + b_j + c_{ij} + \epsilon_{ijk}.$$

Here we have both a fixed effect (for questions, i) and a random effect (for candidates, j). With both fixed and random effects, we speak of a *mixed* model; see §9.1.

With only random effects, we have a *random effects model*, and use notation such as

$$y_{ijk} = \mu + a_i + b_j + c_{ij} + \epsilon_{ijk}.$$

We restrict for simplicity here to the model with no interaction terms:

$$y_{ijk} = \mu + a_i + b_j + \epsilon_{ijk}.$$

Assuming independence of the random variables on the right, the variances add (see e.g. Haigh (2002), Cor. 5.6):

$$\sigma_y^2 = \sigma_a^2 + \sigma_b^2 + \sigma_c^2,$$

in an obvious notation. The terms on the right are called *variance components*; see e.g. Searle, Casella and McCulloch (1992) for a detailed treatment.

Variance components can be traced back to work of Airy in 1861 on astronomical observations (recall that astronomy also led to the development of Least Squares by Legendre and Gauss).

EXERCISES

2.1. (i) Show that if X, Y are positive random variables with joint density $f(x, y)$ their quotient $Z := X/Y$ has density

$$h(z) = \int_0^\infty y f(yz, y) \, dy \quad (z > 0).$$

So if X, Y are independent with densities f, g,

$$h(z) = \int_0^\infty y f(yz) g(y) \, dy \quad (z > 0).$$

(ii) If X has density f and $c > 0$, show that X/c has density

$$f_{X/c}(x) = c f(cx).$$

(iii) Deduce that the Fisher F-distribution $F(m, n)$ has density

$$h(z) = m^{\frac{1}{2}m} n^{\frac{1}{2}n} \frac{\Gamma(\frac{1}{2}m + \frac{1}{2}n)}{\Gamma(\frac{1}{2}m)\Gamma(\frac{1}{2}n)} \cdot \frac{z^{\frac{1}{2}m-1}}{(n + mz)^{\frac{1}{2}(m+n)}} \quad (z > 0).$$

2.2. Using tables or S-Plus/R® produce bounds or calculate the exact probabilities for the following statements. [Note. In S-Plus/R® the command `pf` may prove useful.]
(i) $P(X < 1.4)$ where $X \sim F_{5,17}$,
(ii) $P(X > 1)$ where $X \sim F_{1,16}$,
(iii) $P(X < 4)$ where $X \sim F_{1,3}$,
(iv) $P(X > 3.4)$ where $X \sim F_{19,4}$,
(v) $P(\ln X > -1.4)$ where $X \sim F_{10,4}$.

Fat 1	Fat 2	Fat 3	Fat 4
164	178	175	155
172	191	193	166
168	197	178	149
177	182	171	164
156	185	163	170
195	177	176	168

Table 2.10 Data for Exercise 2.3.

2.3. *Doughnut data.* Doughnuts absorb fat during cooking. The following experiment was conceived to test whether the amount of fat absorbed depends on the type of fat used. Table 2.10 gives the amount of fat absorbed per batch of doughnuts. Produce the one-way Analysis of Variance table for these data. What is your conclusion?

2.4. The data in Table 2.11 come from an experiment where growth is measured and compared to the variable *photoperiod* which indicates the length of daily exposure to light. Produce the one-way ANOVA table for these data and determine whether or not growth is affected by the length of daily light exposure.

Very short	Short	Long	Very long
2	3	3	4
3	4	5	6
1	2	1	2
1	1	2	2
2	2	2	2
1	1	2	3

Table 2.11 Data for Exercise 2.4

2.5. *Unpaired t-test with equal variances.* Under the null hypothesis the statistic t defined as

$$t = \sqrt{\frac{n_1 n_2}{n_1 + n_2}} \, \frac{\left(\overline{X}_1 - \overline{X}_2 - (\mu_1 - \mu_2)\right)}{s}$$

should follow a t distribution with $n_1 + n_2 - 2$ degrees of freedom, where n_1 and n_2 denote the number of observations from samples 1 and 2 and s is the pooled estimate given by

$$s^2 = \frac{(n_1 - 1)s_1^2 + (n_2 - 1)s_2^2}{n_1 + n_2 - 2},$$

where

$$s_1^2 = \frac{1}{n_1 - 1}(\sum x_1^2 - (n_1 - 1)\bar{x}_1^2),$$

$$s_2^2 = \frac{1}{n_2 - 1}(\sum x_2^2 - (n_2 - 1)\bar{x}_2^2).$$

(i) Give the relevant statistic for a test of the hypothesis $\mu_1 = \mu_2$ and $n_1 = n_2 = n$.

(ii) Show that if $n_1 = n_2 = n$ then one-way ANOVA recovers the same results as the unpaired t-test. [Hint. Show that the F-statistic satisfies $F_{1,2(n-1)} = t^2_{2(n-1)}$.]

2.6. Let Y_1, Y_2 be iid $N(0, 1)$. Give values of a and b such that

$$a(Y_1 - Y_2)^2 + b(Y_1 + Y_2)^2 \sim \chi_2^2.$$

2.7. Let Y_1, Y_2, Y_3 be iid $N(0, 1)$. Show that

$$\frac{1}{3}\left[(Y_1 - Y_2)^2 + (Y_2 - Y_3)^2 + (Y_3 - Y_1)^2\right] \sim \chi_2^2.$$

Generalise the above result for a sample $Y_1, Y_2, \ldots, Y_n$ of size n.

2.8. The data in Table 2.12 come from an experiment testing the number of failures out of 100 planted soyabean seeds, comparing four different seed treatments, with no treatment ('check'). Produce the two-way ANOVA table for this data and interpret the results. (We will return to this example in Chapter 8.)

Treatment	Rep 1	Rep 2	Rep 3	Rep 4	Rep 5
Check	8	10	12	13	11
Arasan	2	6	7	11	5
Spergon	4	10	9	8	10
Semesan, Jr	3	5	9	10	6
Fermate	9	7	5	5	3

Table 2.12 Data for Exercise 2.8

2.9. *Photoperiod example revisited.* When we add in knowledge of plant genotype the full data set is as shown in Table 2.13. Produce the two-way ANOVA table and revise any conclusions from Exercise 2.4 in the light of these new data as appropriate.

Genotype	Very short	Short	Long	Very Long
A	2	3	3	4
B	3	4	5	6
C	1	2	1	2
D	1	1	2	2
E	2	2	2	2
F	1	1	2	3

Table 2.13 Data for Exercise 2.9

2.10. *Two-way ANOVA with interactions.* Three varieties of potato are planted on three plots at each of four locations. The yields in bushels are given in Table 2.14. Produce the ANOVA table for these data. Does the interaction term appear necessary? Describe your conclusions.

Variety	Location 1	Location 2	Location 3	Location 4
A	15, 19, 22	17, 10, 13	9, 12, 6	14, 8, 11
B	20, 24, 18	24, 18, 22	12, 15, 10	21, 16, 14
C	22, 17, 14	26, 19, 21	10, 5, 8	19, 15, 12

Table 2.14 Data for Exercise 2.10

2.11. *Two-way ANOVA with interactions.* The data in Table 2.15 give the gains in weight of male rats from diets with different sources and different levels of protein. Produce the two-way ANOVA table with interactions for these data. Test for the presence of interactions between source and level of protein and state any conclusions that you reach.

Source	High Protein	Low Protein
Beef	73, 102, 118, 104, 81,	90, 76, 90, 64, 86,
	107, 100, 87, 117, 111	51, 72, 90, 95, 78
Cereal	98, 74, 56, 111, 95,	107, 95, 97, 80, 98,
	88, 82, 77, 86, 92	74, 74, 67, 89, 58
Pork	94, 79, 96, 98, 102,	49, 82, 73, 86, 81,
	102, 108, 91, 120, 105	97, 106, 70, 61, 82

Table 2.15 Data for Exercise 2.11

3
Multiple Regression

3.1 The Normal Equations

We saw in Chapter 1 how the model

$$y_i = a + bx_i + \epsilon_i, \qquad \epsilon_i \quad \text{iid} \quad N(0, \sigma^2)$$

for simple linear regression occurs. We saw also that we may need to consider two or more regressors. We dealt with two regressors u and v, and could deal with three regressors u, v and w similarly. But in general we will need to be able to handle any number of regressors, and rather than rely on the finite resources of the alphabet it is better to switch to suffix notation, and use the language of vectors and matrices. For a random vector $\mathbf{X}$, we will write $E\mathbf{X}$ for its *mean vector* (thus the mean of the ith coordinate X_i is $E(X_i) = (E\mathbf{X})_i$), and var$(\mathbf{X})$ for its *covariance matrix* (whose (i, j) entry is cov(X_i, X_j)). We will use p regressors, called $x_1, \ldots, x_p$, each with a corresponding parameter $\beta_1, \ldots, \beta_p$ ('p for parameter'). In the equation above, regard a as short for $a.1$, with 1 as a regressor corresponding to a constant term (the intercept term in the context of linear regression). Then for one reading ('a sample of size 1') we have the model

$$y = \beta_1 x_1 + \ldots + \beta_p x_p + \epsilon, \qquad \epsilon_i \quad \sim \quad N(0, \sigma^2).$$

In the general case of a sample of size n, we need two suffices, giving the model equations

$$y_i = \beta_1 x_{i1} + \ldots + \beta_p x_{ip} + \epsilon_i, \qquad \epsilon_i \quad \text{iid} \quad N(0, \sigma^2) \qquad (i = 1, \ldots, n).$$

N.H. Bingham and J.M. Fry, *Regression: Linear Models in Statistics*,
Springer Undergraduate Mathematics Series, DOI 10.1007/978-1-84882-969-5_3,
© Springer-Verlag London Limited 2010

Writing the typical term on the right as $x_{ij}\beta_j$, we recognise the form of a matrix product. Form $y_1, \ldots, y_n$ into a column vector $\mathbf{y}$, $\epsilon_1, \ldots, \epsilon_n$ into a column vector ϵ, $\beta_1, \ldots, \beta_p$ into a column vector β, and x_{ij} into a matrix X (thus $\mathbf{y}$ and ϵ are $n \times 1$, β is $p \times 1$ and X is $n \times p$). Then our system of equations becomes one matrix equation, the *model equation*

$$\mathbf{y} = X\beta + \epsilon. \qquad (ME)$$

This matrix equation, and its consequences, are the object of study in this chapter. Recall that, as in Chapter 1, n is the sample size – the larger the better – while p, the number of parameters, is small – as small as will suffice. We will have more to say on choice of p later. Typically, however, p will be at most five or six, while n could be some tens or hundreds. Thus we must expect n to be *much larger* than p, which we write as

$$n >> p.$$

In particular, the $n \times p$ matrix X has no hope of being invertible, as it is not even square (a common student howler).

Note 3.1

We pause to introduce the objects in the model equation (ME) by name. On the left is $\mathbf{y}$, the *data*, or *response vector*. The last term ϵ is the *error* or *error vector*; β is the *parameter* or *parameter vector*. Matrix X is called the *design matrix*. Although its (i, j) entry arose above as the ith value of the jth regressor, for most purposes from now on x_{ij} is just a *constant*. Emphasis shifts from these constants to the *parameters*, β_j.

Note 3.2

To underline this shift of emphasis, it is often useful to change notation and write A for X, when the model equation becomes

$$\mathbf{y} = A\beta + \epsilon. \qquad (ME)$$

Lest this be thought a trivial matter, we mention that Design of Experiments (initiated by Fisher) is a subject in its own right, on which numerous books have been written, and to which we return in §9.3.

   We will feel free to use either notation as seems most convenient at the time. While X is the natural choice for straight regression problems, as in this chapter, it is less suitable in the general Linear Model, which includes related contexts such as Analysis of Variance (Chapter 2) and Analysis of Covariance (Chapter 5). Accordingly, we shall usually prefer A to X for use in developing theory.

We make a further notational change. As we shall be dealing from now on with vectors rather than scalars, there is no need to remind the reader of this by using boldface type. We may thus lighten the notation by using y for $\mathbf{y}$, etc.; thus we now have

$$y = A\beta + \epsilon, \qquad\qquad (ME)$$

for use in this chapter (in Chapter 4 below, where we again use x as a scalar variable, we use $\mathbf{x}$ for a vector variable).

From the model equation

$$y_i = \sum_{j=1}^{p} a_{ij}\beta_j + \epsilon_i, \qquad \epsilon_i \quad \text{iid} \quad N(0, \sigma^2),$$

the likelihood is

$$
\begin{aligned}
L &= \frac{1}{\sigma^n (2\pi)^{\frac{1}{2}n}} \prod_{i=1}^{n} \exp\left\{ -\frac{1}{2}\left(y_i - \sum_{j=1}^{p} a_{ij}\beta_j\right)^2 / \sigma^2 \right\} \\
&= \frac{1}{\sigma^n (2\pi)^{\frac{1}{2}n}} \exp\left\{ -\frac{1}{2}\sum_{i=1}^{n}\left(y_i - \sum_{j=1}^{p} a_{ij}\beta_j\right)^2 / \sigma^2 \right\},
\end{aligned}
$$

and the log-likelihood is

$$\ell := \log L = \text{const} - n\log\sigma - \frac{1}{2}\left[\sum_{i=1}^{n}\left(y_i - \sum_{j=1}^{p} a_{ij}\beta_j\right)^2\right] / \sigma^2.$$

As before, we use Fisher's Method of Maximum Likelihood, and maximise with respect to β_r: $\partial\ell/\partial\beta_r = 0$ gives

$$\sum_{i=1}^{n} a_{ir}\left(y_i - \sum_{j=1}^{p} a_{ij}\beta_j\right) = 0 \qquad (r = 1, \ldots, p),$$

or

$$\sum_{j=1}^{p}\left(\sum_{i=1}^{n} a_{ir}a_{ij}\right)\beta_j = \sum_{i=1}^{n} a_{ir}y_i.$$

Write $C = (c_{ij})$ for the $p \times p$ matrix

$$C := A^T A,$$

(called the *information matrix* – see Definition 3.10 below), which we note is *symmetric*: $C^T = C$. Then

$$c_{ij} = \sum_{k=1}^{n} (A^T)_{ik} A_{kj} = \sum_{k=1}^{n} a_{ki}a_{kj}.$$

So this says

$$\sum_{j=1}^{p} c_{rj}\beta_j = \sum_{i=1}^{n} a_{ir}y_i = \sum_{i=1}^{n} (A^T)_{ri}y_i.$$

In matrix notation, this is

$$(C\beta)_r = (A^T y)_r \qquad (r = 1, \ldots, p),$$

or combining,

$$C\beta = A^T y, \qquad C := A^T A. \tag{NE}$$

These are the *normal equations*, the analogues for the general case of the normal equations obtained in Chapter 1 for the cases of one and two regressors.

3.2 Solution of the Normal Equations

Our next task is to solve the normal equations for β. Before doing so, we need to check that there exists a unique solution, the condition for which is, from Linear Algebra, that the information matrix $C := A^T A$ should be non-singular (see e.g. Blyth and Robertson (2002a), Ch. 4). This imposes an important condition on the design matrix A. Recall that the *rank* of a matrix is the maximal number of independent rows or columns. If this is as big as it could be given the size of the matrix, the matrix is said to have *full rank*, otherwise it has *deficient rank*. Since A is $n \times p$ with $n >> p$, A has full rank if its rank is p.

Recall from Linear Algebra that a square matrix C is *non-negative definite* if

$$x^T C x \geq 0$$

for all vectors x, while C is *positive definite* if

$$x^T C x > 0 \qquad \forall x \neq 0$$

(see e.g. Blyth and Robertson (2002b), Ch. 8). A positive definite matrix is non-singular, so invertible; a non-negative definite matrix need not be.

Lemma 3.3

If A $(n \times p, n > p)$ has full rank p, $C := A^T A$ is positive definite.

Proof

As A has full rank, there is no vector x with $Ax = 0$ other than the zero vector (such an equation would give a non-trivial linear dependence relation between the columns of A). So

$$(Ax)^T Ax = x^T A^T Ax = x^T C x = 0$$

only for $x = 0$, and is > 0 otherwise. This says that C is positive definite, as required.   □

Note 3.4

The same proof shows that $C := A^T A$ is always non-negative definite, regardless of the rank of A.

Theorem 3.5

For A full rank, the normal equations have the unique solution

$$\hat{\beta} = C^{-1} A^T y = (A^T A)^{-1} A^T y. \qquad (\hat{\beta})$$

Proof

In the full-rank case, C is positive definite by Lemma 3.3, so invertible, so we may solve the normal equations to obtain the solution above.   □

From now on, we restrict attention to the full-rank case: the design matrix A, which is $n \times p$, has full rank p.

Note 3.6

The distinction between the full- and deficient-rank cases is the same as that between the general and singular cases that we encountered in Chapter 1 in connection with the bivariate normal distribution. We will encounter it again later in Chapter 4, in connection with the multivariate normal distribution. In fact, this distinction bedevils the whole subject. Linear dependence causes rank-deficiency, in which case we should identify the linear dependence relation, use it to express some regressors (or columns of the design matrix) in terms of others, eliminate the redundant regressors or columns, and begin again in a lower dimension, where the problem will have full rank. What is worse is that *near-linear dependence* – which when regressors are at all numerous is not uncommon – means that one is close to rank-deficiency, and this makes things numerically unstable. Remember that in practice, we work numerically, and when one is within rounding error of rank-deficiency, one is close to disaster. We shall return to this vexed matter later (§4.4), in connection with *multicollinearity*. We note in passing that Numerical Linear Algebra is a subject in its own right; for a monograph treatment, see e.g. Golub and Van Loan (1996).

Just as in Chapter 1, the functional form of the normal likelihood means that maximising the likelihood minimises the sum of squares

$$SS := (y - A\beta)^T (y - A\beta) = \sum_{i=1}^{n} \left(y_i - \sum_{j=1}^{p} a_{ij}\beta_j \right)^2.$$

Accordingly, we have as before the following theorem.

Theorem 3.7

The solutions $(\hat{\beta})$ to the normal equations (NE) are both the maximum-likelihood estimators and the least-squares estimators of the parameters β.

There remains the task of estimating the remaining parameter σ. At the maximum, $\beta = \hat{\beta}$. So taking $\partial SS / \partial \sigma = 0$ in the log-likelihood

$$\ell := \log L = \text{const} - n \log \sigma - \frac{1}{2} \left[\sum_{i=1}^{n} \left(y_i - \sum_{j=1}^{p} a_{ij}\beta_j \right)^2 \right] / \sigma^2$$

gives, at the maximum,

$$-\frac{n}{\sigma} + \frac{1}{\sigma^3} \sum_{i=1}^{n} \left(y_i - \sum_{j=1}^{p} a_{ij}\beta_j \right)^2 = 0.$$

At the maximum, $\beta = \hat{\beta}$; rearranging, we have at the maximum that

$$\sigma^2 = \frac{1}{n} \sum_{i=1}^{n} \left(y_i - \sum_{j=1}^{p} a_{ij}\hat{\beta}_j \right)^2.$$

This sum of squares is, by construction, the minimum value of the total sum of squares SS as the parameter β varies, the minimum being attained at the least-squares estimate $\hat{\beta}$. This minimised sum of squares is called the *sum of squares for error*, SSE:

$$SSE = \sum_{i=1}^{n} \left(y_i - \sum_{j=1}^{p} a_{ij}\hat{\beta}_j \right)^2 = \left(y - A\hat{\beta} \right)^T \left(y - A\hat{\beta} \right),$$

so-called because, as we shall see in Corollary 3.23 below, the unbiased estimator of the error variance σ^2 is $\hat{\sigma}^2 = SSE/(n - p)$.

We call

$$\hat{y} := A\hat{\beta}$$

the *fitted values*, and

$$e := y - \hat{y},$$

the difference between the actual values (data) and fitted values, the *residual vector*. If $e = (e_1, \ldots, e_n)$, the e_i are the *residuals*, and the sum of squares for error

$$SSE = \sum_{i=1}^{n} (y_i - \hat{y}_i)^2 = \sum_{i=1}^{n} e_i^2$$

is the *sum of squared residuals*.

Note 3.8

We pause to discuss unbiasedness and degrees of freedom (df). In a first course in Statistics, one finds the maximum-likelihood estimators (MLEs) $\hat{\mu}$, $\hat{\sigma}^2$ of the parameters μ, σ^2 in a normal distribution $N(\mu, \sigma^2)$. One finds

$$\hat{\mu} = \bar{x}, \qquad \hat{\sigma}^2 = s_x^2 := \frac{1}{n}\sum_{i=1}^{n}(x_i - \bar{x})^2$$

(and the distributions are given by $\bar{x} \sim N(\mu, \sigma^2/n)$ and $n\hat{\sigma}^2/\sigma^2 \sim \chi^2(n-1)$). But this is a *biased* estimator of σ^2; to get an *unbiased* estimator, one has to replace n in the denominator above by $n-1$ (in distributional terms: the mean of a chi-square is its df). This is why many authors use $n-1$ in place of n in the denominator when they *define* the sample variance (and we warned, when we used n in Chapter 1, that this was not universal!), giving what we will call the *unbiased* sample variance,

$$s_u^2 := \frac{1}{(n-1)}\sum_{i=1}^{n}(x_i - \bar{x})^2.$$

The problem is that to estimate σ^2, one has first to estimate μ by $\bar{x}$. *Every time one has to estimate a parameter from the data, one loses a degree of freedom.* In this one-dimensional problem, the df accordingly decreases from n to $n-1$.

Returning to the general case: here we have to estimate p parameters, $\beta_1, \ldots, \beta_p$. Accordingly, we lose p degrees of freedom, and to get an unbiased estimator we have to divide, not by n as above but by $n-p$, giving the estimator

$$\hat{\sigma}^2 = \frac{1}{(n-p)}SSE.$$

Since n is much larger than p, the difference between this (unbiased) estimator and the previous (maximum-likelihood) version is not large, but it is worthwhile, and so we shall work with the unbiased version unless otherwise stated. We find its distribution in §3.4 below (and check it is unbiased – Corollary 3.23).

Note 3.9 (Degrees of Freedom)

Recall that n is our sample size, that p is our number of parameters, and that n is much greater than p. The need to estimate p parameters, which reduces the degrees of freedom from n to $n-p$, thus effectively reduces the sample size by this amount. We can think of the degrees of freedom as a measure of the *amount of information available* to us.

This interpretation is in the minds of statisticians when they prefer one procedure to another because it 'makes more degrees of freedom available' for the task in hand. We should always keep the degrees of freedom of all relevant terms (typically, sums of squares, or quadratic forms in normal variates) in mind, and think of keeping this large as being desirable.

We rewrite our conclusions so far in matrix notation. The total sum of squares is

$$SS := \sum_{i=1}^{n} \left(y_i - \sum_{j=1}^{p} a_{ij}\beta_j \right)^2 = (y - A\beta)^T (y - A\beta);$$

its minimum value with respect to variation in β is the sum of squares for error

$$SSE = \sum_{i=1}^{n} \left(y_i - \sum_{j=1}^{p} a_{ij}\hat{\beta}_j \right)^2 = \left(y - A\hat{\beta} \right)^T \left(y - A\hat{\beta} \right),$$

where $\hat{\beta}$ is the solution to the normal equations (NE). Note that SSE is a *statistic* – we can calculate it from the data y and $\hat{\beta} = C^{-1}A^T y$, unlike SS which contains unknown parameters β.

One feature is amply clear already. To carry through a regression analysis in practice, we must perform considerable matrix algebra – or, with actual data, numerical matrix algebra – involving in particular the inversion of the $p \times p$ matrix $C := A^T A$. With matrices of any size, the calculations may well be laborious to carry out by hand. In particular, *matrix inversion* to find C^{-1} will be unpleasant for matrices larger than 2×2, even though C – being symmetric and positive definite – has good properties. For matrices of any size, one needs computer assistance. The package MATLAB®[1] is specially designed with matrix operations in mind. General mathematics packages such as Mathematica®[2] or Maple®[3] have a matrix inversion facility; so too do a number of statistical packages – for example, the `solve` command in S-Plus/R®.

QR Decomposition

The numerical solution of the normal equations $((NE)$ in §3.1, $(\hat{\beta})$ in Theorem 3.5) is simplified if the design matrix A (which is $n \times p$, and of full rank p) is given its *QR decomposition*

$$A = QR,$$

where Q is $n \times p$ and has *orthonormal columns* – so

$$Q^T Q = I$$

[1] MATLAB®, Simulink® and Symbolic Math Toolbox™ are trademarks of The MathWorks, Inc., 3 Apple Hill Drive, Natick, MA, 01760-2098, USA, http://www. mathworks.com

[2] Mathematica® is a registered trademark of Wolfram Research, Inc., 100 Trade Center Drive, Champaign, IL 61820-7237, USA, http://www.wolfram.com

[3] Maple™ is a trademark of Waterloo Maple Inc., 615 Kumpf Drive, Waterloo, Ontario, Canada N2V 1K8, http://www.maplesoft.com

– and R is $p \times p$, *upper triangular*, and non-singular (has no zeros on the diagonal). This is always possible; see below. The normal equations $A^T A \hat{\beta} = A^T y$ then become

$$R^T Q^T Q R \hat{\beta} = R^T Q^T y,$$

or

$$R^T R \hat{\beta} = R^T Q^T y,$$

as $Q^T Q = I$, or

$$R \hat{\beta} = Q^T y,$$

as R, and so also R^T, is non-singular. This system of linear equations for $\hat{\beta}$ has an upper triangular matrix R, and so may be solved simply by back-substitution, starting with the bottom equation and working upwards.

The QR decomposition is just the expression in matrix form of the process of *Gram–Schmidt orthogonalisation*, for which see e.g. Blyth and Robertson (2002b), Th. 1.4. Write A as a row of its columns,

$$A = (a_1, \ldots, a_p);$$

the n-vectors a_i are linearly independent as A has full rank p. Write $q_1 := a_1 / \|a_1\|$, and for $j = 2, \ldots, p$,

$$q_j := w_j / \|w_j\|, \qquad \text{where} \qquad w_j := a_j - \sum_{k=1}^{j-1} (a_j^T q_k) q_k.$$

Then the q_j are orthonormal (are mutually orthogonal unit vectors), which span the column-space of A (Gram-Schmidt orthogonalisation is this process of passing from the a_j to the q_j). Each q_j is a linear combination of $a_1, \ldots, a_j$, and the construction ensures that, conversely, each a_j is a linear combination of $q_1, \ldots, q_j$. That is, there are scalars r_{kj} with

$$a_j = \sum_{k=1}^{j} r_{kj} q_k \qquad (j = 1, \ldots, p).$$

Put $r_{kj} = 0$ for $k > j$. Then assembling the p columns a_j into the matrix A as above, this equation becomes

$$A = QR,$$

as required.

Note 3.10

Though useful as a theoretical tool, the Gram–Schmidt orthogonalisation process is not numerically stable. For numerical implementation, one needs a stable variant, the modified Gram-Schmidt process. For details, see Golub and Van Loan (1996), §5.2. They also give other forms of the QR decomposition (Householder, Givens, Hessenberg etc.).

3.3 Properties of Least-Squares Estimators

We have assumed *normal errors* in our model equations, (ME) of §3.1. But (until we need to assume normal errors in §3.5.2), we may work more generally, and assume only

$$Ey = A\beta, \qquad \text{var}(y) = \sigma^2 I. \qquad (ME^*)$$

We must then restrict ourselves to the Method of Least Squares, as without distributional assumptions we have no likelihood function, so cannot use the Method of Maximum Likelihood.

Linearity. The least-squares estimator

$$\hat{\beta} = C^{-1} A^T y$$

is *linear* in the data y.

Unbiasedness.

$$E\hat{\beta} = C^{-1} A^T Ey = C^{-1} A^T A\beta = C^{-1} C\beta = \beta :$$

$\hat{\beta}$ is an unbiased estimator of β.

Covariance matrix.

$$\begin{aligned}
\text{var}(\hat{\beta}) = \text{var}(C^{-1} A^T y) &= C^{-1} A^T (\text{var}(y))(C^{-1} A^T)^T \\
&= C^{-1} A^T . \sigma^2 I . AC^{-1} \qquad (C = C^T) \\
&= \sigma^2 . C^{-1} A^T . AC^{-1} \\
&= \sigma^2 C^{-1} \qquad (C = A^T A).
\end{aligned}$$

We wish to keep the variances of our estimators of our p parameters β_i small, and these are the diagonal elements of the covariance matrix above; similarly for the covariances (off-diagonal elements). The smaller the variances, the more precise our estimates, and the more information we have. This motivates the next definition.

Definition 3.11

The matrix $C := A^T A$, with A the design matrix, is called the *information matrix*.

Note 3.12

1. The variance σ^2 in our errors ϵ_i (which we of course wish to keep small) is usually beyond our control. However, at least at the stage of design and planning of the experiment, the design matrix A may well be within our control; hence so will be the information matrix $C := A^T A$, which we wish to maximise (in some sense), and hence so will be C^{-1}, which we wish to minimise in some sense. We return to this in §9.3 in connection with Design of Experiments.

2. The term information matrix is due to Fisher. It is also used in the context of parameter estimation by the *method of maximum likelihood*. One has the likelihood $L(\theta)$, with θ a vector parameter, and the log-likelihood $\ell(\theta) := \log L(\theta)$. The information matrix is the negative of the Hessian (matrix of second derivatives) of the log-likelihood: $I(\theta) := (I_{ij}(\theta))_{i,j=1}^p$, when

$$I_{ij}(\theta) := -\frac{\partial^2}{\partial \theta_i \partial \theta_j}\ell(\theta).$$

Under suitable regularity conditions, the maximum likelihood estimator $\hat{\theta}$ is asymptotically normal and unbiased, with variance matrix $(nI(\theta))^{-1}$; see e.g. Rao (1973), 5a.3, or Cramér (1946), §33.3.

Unbiased linear estimators.   Now let $\tilde{\beta} := By$ be *any* unbiased linear estimator of β (B a $p \times n$ matrix). Then

$$E\tilde{\beta} = BEy = BA\beta = \beta$$

– and so $\tilde{\beta}$ is an unbiased estimator for β – iff

$$BA = I.$$

Note that

$$\mathrm{var}(\tilde{\beta}) = B\mathrm{var}(y)B^T = B.\sigma^2 I.B^T = \sigma^2 BB^T.$$

In the context of linear regression, as here, it makes sense to restrict attention to linear estimators. The two most obviously desirable properties of such estimators are unbiasedness (to get the mean right), and being minimum variance (to get maximum precision). An estimator with both these desirable properties may be termed a *best estimator*. A linear one is then a best linear unbiased estimator or BLUE (such acronyms are common in Statistics, and useful; an alternative usage is minimum variance unbiased linear estimate, or MVULE, but this is longer and harder to say). It is remarkable that the least-squares estimator that we have used above is best in this sense, or BLUE.

Theorem 3.13 (Gauss–Markov Theorem)

Among all unbiased linear estimators $\tilde{\beta} = By$ of β, the least-squares estimator $\hat{\beta} = C^{-1}A^T y$ has the minimum variance in each component. That is $\hat{\beta}$ is the BLUE.

Proof

By above, the covariance matrix of an arbitrary unbiased linear estimate $\tilde{\beta} = By$ and of the least-squares estimator $\hat{\beta}$ are given by

$$\text{var}(\tilde{\beta}) = \sigma^2 BB^T \quad \text{and} \quad \text{var}(\hat{\beta}) = \sigma^2 C^{-1}.$$

Their difference (which we wish to show is non-negative) is

$$\text{var}(\tilde{\beta}) - \text{var}(\hat{\beta}) = \sigma^2[BB^T - C^{-1}].$$

Now using symmetry of C, C^{-1}, and $BA = I$ (so $A^T B^T = I$) from above,

$$(B - C^{-1}A^T)(B - C^{-1}A^T)^T = (B - C^{-1}A^T)(B^T - AC^{-1}).$$

Further,

$$
\begin{aligned}
(B - C^{-1}A^T)(B^T - AC^{-1}) &= BB^T - BAC^{-1} - C^{-1}A^T B^T + C^{-1}A^T AC^{-1} \\
&= BB^T - C^{-1} - C^{-1} + C^{-1} \quad (C = A^T A) \\
&= BB^T - C^{-1}.
\end{aligned}
$$

Combining,

$$\text{var}(\tilde{\beta}) - \text{var}(\hat{\beta}) = \sigma^2(B - C^{-1}A^T)(B - C^{-1}A^T)^T.$$

Now for a matrix $M = (m_{ij})$,

$$(MM^T)_{ii} = \sum_k m_{ik}(M^T)_{ki} = \sum_k m_{ik}^2,$$

the sum of the squares of the elements on the ith row of matrix M. So the ith diagonal entry above is

$$\text{var}(\tilde{\beta}_i) = \text{var}(\hat{\beta}_i) + \sigma^2(\text{sum of squares of elements on } i\text{th row of } B - C^{-1}A^T).$$

So

$$\text{var}(\tilde{\beta}_i) \geq \text{var}(\hat{\beta}_i),$$

and

$$\text{var}(\tilde{\beta}_i) = \text{var}(\hat{\beta}_i)$$

iff $B - C^{-1}A^T$ has ith row zero. So *some* $\tilde{\beta}_i$ has greater variance than $\hat{\beta}_i$ unless $B = C^{-1}A^T$ (i.e., unless *all* rows of $B - C^{-1}A^T$ are zero) – that is, unless $\tilde{\beta} = By = C^{-1}A^T y = \hat{\beta}$, the least-squares estimator, as required.   □

One may summarise all this as: whether or not errors are assumed normal,
LEAST SQUARES IS BEST.

Note 3.14

The Gauss–Markov theorem is in fact a misnomer. It is due to Gauss, in the early eighteenth century; it was treated in the book Markov (1912) by A. A. Markov (1856–1922). A misreading of Markov's book gave rise to the impression that he had rediscovered the result, and the name Gauss–Markov theorem has stuck (partly because it is useful!).

Estimability. A linear combination $c^T \beta = \sum_{i=1}^{p} c_i \beta_i$, with $c = (c_1, \ldots, c_p)^T$ a known p-vector, is called *estimable* if it has an unbiased linear estimator, $b^T y = \sum_{i=1}^{n} b_i y_i$, with $b = (b_1, \ldots, b_n)^T$ a known n-vector. Then

$$E(b^T y) = b^T E(y) = b^T A \beta = c^T \beta.$$

This can hold identically in the unknown parameter β iff

$$c^T = b^T A,$$

that is, c is a linear combination (by the n-vector b) of the n rows (p-vectors) of the design matrix A. The concept is due to R. C. Bose (1901–1987) in 1944.

In the full-rank case considered here, the rows of A span a space of full dimension p, and so all linear combinations are estimable. But in the defective rank case with rank $k < p$, the estimable functions span a space of dimension k, and non-estimable linear combinations exist.

3.4 Sum-of-Squares Decompositions

We define the *sum of squares for regression, SSR*, by

$$SSR := (\hat{\beta} - \beta)^T C (\hat{\beta} - \beta).$$

Since this is a quadratic form with matrix C which is positive definite, we have $SSR \geq 0$, and $SSR > 0$ unless $\hat{\beta} = \beta$, that is, unless the least-squares estimator is exactly right (which will, of course, never happen in practice).

Theorem 3.15 (Sum-of-Squares Decomposition)

$$SS = SSR + SSE. \tag{SSD}$$

Proof

Write

$$y - A\beta = (y - A\hat{\beta}) + A(\hat{\beta} - \beta).$$

Now multiply the vector on each side by its transpose (that is, form the sum of squares of the coordinates of each vector). On the left, we obtain

$$SS = (y - A\beta)^T (y - A\beta),$$

the total sum of squares. On the right, we obtain three terms. The first squared term is

$$SSE = (y - A\hat{\beta})^T (y - A\hat{\beta}),$$

the sum of squares for error. The second squared term is

$$(A(\hat{\beta} - \beta))^T A(\hat{\beta} - \beta) = (\hat{\beta} - \beta)^T A^T A(\hat{\beta} - \beta) = (\hat{\beta} - \beta)^T C(\hat{\beta} - \beta) = SSR,$$

the sum of squares for regression. The cross terms on the right are

$$(y - A\hat{\beta})^T A(\hat{\beta} - \beta)$$

and its transpose, which are the same as both are scalars. But

$$A^T (y - A\hat{\beta}) = A^T y - A^T A\hat{\beta} = A^T y - C\hat{b} = 0,$$

by the normal equations (NE) of §3.1-3.2. Transposing,

$$(y - A\hat{\beta})^T A = 0.$$

So both cross terms vanish, giving $SS = SSR + SSE$, as required.   □

Corollary 3.16

We have that

$$SSE = \min_{\beta} SS,$$

the minimum being attained at the least-squares estimator $\hat{\beta} = C^{-1}A^T y$.

Proof

$SSR \geq 0$, and $= 0$ iff $\beta = \hat{\beta}$.   □

We now introduce the geometrical language of *projections*, to which we return in e.g. §3.5.3 and §3.6 below. The relevant mathematics comes from Linear Algebra; see the definition below. As we shall see, doing regression with p regressors amounts to an *orthogonal projection* on an appropriate p-dimensional subspace in n-dimensional space. The sum-of-squares decomposition involved can be visualised geometrically as an instance of *Pythagoras's Theorem*, as in the familiar setting of plane or solid geometry.

Definition 3.17

Call a linear transformation $P : V \to V$ a *projection* onto V_1 along V_2 if V is the direct sum $V = V_1 \oplus V_2$, and if $x = (x_1, x_2)^T$ with $Px = x_1$.

Then (Blyth and Robertson (2002b), Ch.2, Halmos (1979), §41) $V_1 = \text{Im } P = \text{Ker } (I - P)$, $V_2 = \text{Ker } P = \text{Im } (I - P)$.

Recall that a square matrix is *idempotent* if it is its own square $M^2 = M$. Then (Halmos (1979), §41), M is idempotent iff it is a projection.

For use throughout the rest of the book, with A the design matrix and $C := A^T A$ the information matrix, we write

$$P := AC^{-1}A^T$$

('P for projection' – see below). We note that P is symmetric. Note also

$$Py = AC^{-1}A^T y = A\hat{\beta},$$

by the normal equations (NE).

Lemma 3.18

P and $I - P$ are idempotent, and so are projections.

Proof

$P^2 = AC^{-1}A^T.AC^{-1}A^T = AC^{-1}A^T = P$:

$$P^2 = P.$$

$$(I - P)^2 = I - 2P + P^2 = I - 2P + P = I - P. \qquad \square$$

We now rewrite the two terms SSR and SSE on the right in Theorem 3.15 in the language of projections. Note that the first expression for SSE below shows again that it is a *statistic* – a function of the data (not involving unknown parameters), and so can be *calculated* from the data.

Theorem 3.19

$$SSE = y^T(I - P)y = (y - A\beta)^T(I - P)(y - A\beta),$$
$$SSR = (y - A\beta)^T P(y - A\beta).$$

Proof

As $SSE := \left(y - A\hat{\beta}\right)^T \left(y - A\hat{\beta}\right)$, and $A\hat{\beta} = Py$,

$$
\begin{aligned}
SSE &= \left(y - A\hat{\beta}\right)^T \left(y - A\hat{\beta}\right) \\
&= (y - Py)^T(y - Py) = y^T(I - P)(I - P)y = y^T(I - P)y,
\end{aligned}
$$

as $I - P$ is a projection.

For SSR, we have that

$$SSR := \left(\hat{\beta} - \beta\right)^T C \left(\hat{\beta} - \beta\right) = \left(\hat{\beta} - \beta\right)^T A^T A \left(\hat{\beta} - \beta\right).$$

But

$$\left(\hat{\beta} - \beta\right) = C^{-1} A^T y - \beta = C^{-1} A^T y - C^{-1} A^T A\beta = C^{-1} A^T(y - A\beta),$$

so

$$
\begin{aligned}
SSR &= (y - A\beta)^T AC^{-1}.A^T A.C^{-1} A^T(y - A\beta) \\
&= (y - A\beta)^T AC^{-1} A^T(y - A\beta) \qquad (A^T A = C) \\
&= (y - A\beta)^T P(y - A\beta),
\end{aligned}
$$

as required. The second formula for SSE follows from this and (SSD) by subtraction. $\qquad\qquad\Box$

Coefficient of Determination

The *coefficient of determination* is defined as R^2, where R is the (sample) correlation coefficient of the data and the fitted values that is of the pairs $(y_i, \hat{y}_i)$:

$$R := \sum (y_i - \bar{y}) (\hat{y}_i - \bar{\hat{y}}) \left/ \sqrt{\sum (y_i - \bar{y})^2 \sum (\hat{y}_i - \bar{\hat{y}})^2}\right..$$

Thus $-1 \leq R \leq 1$, $0 \leq R^2 \leq 1$, and R^2 is a measure of the *goodness of fit* of the fitted values to the data.

Theorem 3.20

$$R^2 = 1 - \frac{SSE}{\sum (y_i - \bar{y})^2}.$$

For reasons of continuity, we postpone the proof to §3.4.1 below. Note that $R^2 = 1$ iff $SSE = 0$, that is, all the residuals are 0, and the fitted values are the exact values. As noted above, we will see in §3.6 that regression (estimating p parameters from n data points) amounts to a *projection* of the n-dimensional data space onto an p-dimensional hyperplane. So $R^2 = 1$ iff the data points lie in an p-dimensional hyperplane (generalising the situation of Chapter 1, where $R^2 = 1$ iff the data points lie on a line). In our full-rank (non-degenerate) case, this will not happen (see Chapter 4 for the theory of the relevant multivariate normal distribution), but the bigger R^2 is (or the smaller SSE is), the better the fit of our regression model to the data.

Note 3.21

R^2 provides a useful summary of the proportion of the variation in a data set explained by a regression. However, as discussed in Chapters 5 and 11 of Draper and Smith (1998) high values of R^2 can be misleading. In particular, we note that the values R^2 will tend to increase as additional terms are added to the model, irrespective of whether those terms are actually needed. An adjusted R^2 statistic which adds a penalty to complex models can be defined as

$$R_a^2 = 1 - (1 - R^2) \left(\frac{n-1}{n-p} \right),$$

where n is the number of parameters and $n-p$ is the number of residual degrees of freedom; see Exercises 3.3, and §5.2 for a treatment of models penalised for complexity.

We note a result for later use.

Proposition 3.22 (Trace Formula)

$$E(x^T A x) = \text{trace}(A.\text{var}(x)) + Ex^T.A.Ex.$$

Proof

$$x^T A x = \sum_{ij} a_{ij} x_i x_j,$$

so by linearity of E,

$$E[x^T A x] = \sum_{ij} a_{ij} E[x_i x_j].$$

Now $\text{cov}(x_i, x_j) = E(x_i x_j) - (Ex_i)(Ex_j)$, so

$$
\begin{aligned}
E\left[x^T A x\right] &= \sum_{ij} a_{ij} \left[\text{cov}(x_i x_j) + Ex_i.Ex_j\right] \\
&= \sum_{ij} a_{ij}\text{cov}(x_i x_j) + \sum_{ij} a_{ij}.Ex_i.Ex_j.
\end{aligned}
$$

The second term on the right is $Ex^T A E x$. For the first, note that

$$\text{trace}(AB) = \sum_i (AB)_{ii} = \sum_{ij} a_{ij} b_{ji} = \sum_{ij} a_{ij} b_{ij},$$

if B is symmetric. But covariance matrices are symmetric, so the first term on the right is $\text{trace}(A\,\text{var}(x))$, as required.   □

Corollary 3.23

$$\text{trace}(P) = p, \qquad \text{trace}(I - P) = n - p, \qquad E(SSE) = (n - p)\sigma^2.$$

So $\hat{\sigma}^2 := SSE/(n - p)$ is an unbiased estimator for σ^2.

Proof

By Theorem 3.19, SSE is a quadratic form in $y - A\beta$ with matrix $I - P = I - AC^{-1}A^T$. Now

$$\text{trace}(I - P) = \text{trace}(I - AC^{-1}A^T) = \text{trace}(I) - \text{trace}(AC^{-1}A^T).$$

But $\text{trace}(I) = n$ (as here I is the $n \times n$ identity matrix), and as $\text{trace}(AB) = \text{trace}(BA)$ (see Exercise 3.12),

$$\text{trace}(P) = \text{trace}(AC^{-1}A^T) = \text{trace}(C^{-1}A^T A) = \text{trace}(I) = p,$$

as here I is the $p \times p$ identity matrix. So

$$\text{trace}(I - P) = \text{trace}(I - AC^{-1}A^T) = n - p.$$

Since $Ey = A\beta$ and $\text{var}(y) = \sigma^2 I$, the Trace Formula gives

$$E(SSE) = (n - p)\sigma^2.$$

$\square$

This last formula is analogous to the corresponding ANOVA formula $E(SSE) = (n - r)\sigma^2$ of §2.6. In §4.2 we shall bring the subjects of regression and ANOVA together.

3.4.1 Coefficient of determination

We now give the proof of Theorem 3.20, postponed in the above.

Proof

As at the beginning of Chapter 3 we may take our first regressor as 1, corresponding to the intercept term (this is not always present, but since R is translation-invariant, we may add an intercept term without changing R). The first of the normal equations then results from differentiating

$$\sum (y_i - \beta_1 - a_{2i}\beta_2 - \ldots - a_{pi}\beta_p)^2 = 0$$

with respect to β_1, giving

$$\sum (y_i - \beta_1 - a_{2i}\beta_2 - \ldots - a_{pi}\beta_p) = 0.$$

At the minimising values $\hat{\beta}_j$, this says

$$\sum (y_i - \hat{y}_i) = 0.$$

So

$$\bar{y} = \bar{\hat{y}}, \tag{a}$$

and also

$$\begin{aligned}
\sum (y_i - \hat{y}_i)(\hat{y}_i - \bar{y}) &= \sum (y_i - \hat{y}_i)\hat{y}_i \\
&= (y - \hat{y})^T \hat{y} \\
&= (y - Py)^T Py \\
&= y^T (I - P)Py \\
&= y^T (P - P^2)y,
\end{aligned}$$

so

$$\sum(y_i - \hat{y}_i)(\hat{y}_i - \overline{y}) = 0, \qquad (b)$$

as P is a projection. So

$$\sum(y_i - \overline{y})^2 = \sum[(y_i - \hat{y}_i) + (\hat{y}_i - \overline{y})]^2 = \sum(y_i - \hat{y}_i)^2 + \sum(\hat{y}_i - \overline{y})^2, \quad (c)$$

since the cross-term is 0. Also, in the definition of R,

$$\begin{aligned}
\sum(y_i - \overline{y})(\hat{y}_i - \overline{\hat{y}}) &= \sum(y_i - \overline{y})(\hat{y}_i - \overline{y}) \quad \text{(by (a))} \\
&= \sum[(y_i - \hat{y}_i) + (\hat{y}_i - \overline{y})](\hat{y}_i - \overline{y}) \\
&= \sum(\hat{y}_i - \overline{y})^2 \quad \text{(by (b))}.
\end{aligned}$$

So

$$R^2 = \frac{\left[\sum(\hat{y}_i - \overline{y})^2\right]^2}{\left(\sum(y_i - \overline{y})^2 \sum(\hat{y}_i - \overline{y})^2\right)} = \frac{\sum(\hat{y}_i - \overline{y})^2}{\sum(y_i - \overline{y})^2}.$$

By (c),

$$\begin{aligned}
R^2 &= \frac{\sum(\hat{y}_i - \overline{y})^2}{\sum(y_i - \hat{y}_i)^2 + \sum(\hat{y}_i - \overline{y})^2} \\
&= 1 - \frac{\sum(y_i - \hat{y}_i)^2}{\sum(y_i - \hat{y}_i)^2 + \sum(\hat{y}_i - \overline{y})^2} \\
&= 1 - \frac{SSE}{\sum(y_i - \overline{y})^2},
\end{aligned}$$

by (c) again and the definition of SSE. $\qquad\qquad\square$

3.5 Chi-Square Decomposition

Recall (Theorem 2.2) that if $x = x_1, \ldots, x_n$ is $N(0, I)$ – that is, if the x_i are iid $N(0, 1)$ – and we change variables by an orthogonal transformation B to

$$y := Bx,$$

then also $y \sim N(0, I)$. Recall from Linear Algebra (e.g. Blyth and Robertson (2002a) Ch. 9) that λ is an *eigenvalue* of a matrix A with *eigenvector* $x \ (\neq 0)$ if

$$Ax = \lambda x$$

(x is *normalised* if $x^T x = \Sigma_i x_i^2 = 1$, as is always possible).

Recall also (see e.g. Blyth and Robertson (2002b), Corollary to Theorem 8.10) that if A is a real symmetric matrix, then A can be diagonalised by an orthogonal transformation B, to D, say:

$$B^T A B = D$$

(see also Theorem 4.12 below, Spectral Decomposition) and that (see e.g. Blyth and Robertson (2002b), Ch. 9) if λ is an eigenvalue of A,

$$|D - \lambda I| = |B^T A B - \lambda I| = |B^T A B - \lambda B^T B| = |B^T| |A - \lambda I| |B| = 0.$$

Then a quadratic form in normal variables with matrix A is also a quadratic form in normal variables with matrix D, as

$$x^T A x = x^T B D B^T x = y^T D y, \qquad y := B^T x.$$

3.5.1 Idempotence, Trace and Rank

Recall that a (square) matrix M is *idempotent* if $M^2 = M$.

Proposition 3.24

If B is idempotent,

(i) its eigenvalues λ are 0 or 1,

(ii) its trace is its rank.

Proof

(i) If λ is an eigenvalue of B, with eigenvector x, $Bx = \lambda x$ with $x \neq 0$. Then

$$B^2 x = B(Bx) = B(\lambda x) = \lambda(Bx) = \lambda(\lambda x) = \lambda^2 x,$$

so λ^2 is an eigenvalue of B^2 (always true – that is, does not need idempotence). So

$$\lambda x = Bx = B^2 x = \ldots = \lambda^2 x,$$

and as $x \neq 0$, $\lambda = \lambda^2$, $\lambda(\lambda - 1) = 0$: $\lambda = 0$ or 1.

(ii)

$$
\begin{aligned}
\text{trace}(B) &= \text{sum of eigenvalues} \\
&= \text{\# non-zero eigenvalues} \\
&= rank(B). \qquad \square
\end{aligned}
$$

Corollary 3.25

$$rank(P) = p, \qquad rank(I - P) = n - p.$$

Proof

This follows from Corollary 3.23 and Proposition 3.24.   □

Thus $n = p + (n - p)$ is an instance of the Rank–Nullity Theorem ('dim source =dim Ker + dim Im'): Blyth and Robertson (2002a), Theorem 6. 4) applied to P, $I - P$.

3.5.2 Quadratic forms in normal variates

We will be interested in symmetric projection (so idempotent) matrices P. Because their eigenvalues are 0 and 1, we can diagonalise them by orthogonal transformations to a diagonal matrix of 0s and 1s. So if P has rank r, a quadratic form $x^T P x$ can be reduced to a sum of r squares of standard normal variates. By relabelling variables, we can take the 1s to precede the 0s on the diagonal, giving

$$x^T P x = y_1^2 + \ldots + y_r^2, \qquad y_i \quad \text{iid} \quad N(0, \sigma^2).$$

So $x^T P x$ is σ^2 times a $\chi^2(r)$-distributed random variable.

To summarise:

Theorem 3.26

If P is a symmetric projection of rank r and the x_i are independent $N(0, \sigma^2)$, the quadratic form

$$x^T P x \sim \sigma^2 \chi^2(r).$$

3.5.3 Sums of Projections

As we shall see below, a sum-of-squares decomposition, which expresses a sum of squares (chi-square distributed) as a sum of independent sums of squares (also chi-square distributed) corresponds to a decomposition of the identity I

as a sum of orthogonal projections. Thus Theorem 3.13 corresponds to $I = P + (I - P)$, but in Chapter 2 we encountered decompositions with more than two summands (e.g., $SS = SSB + SST + SSI$ has three). We turn now to the general case.

Suppose that $P_1, \ldots, P_k$ are symmetric projection matrices with sum the identity:

$$I = P_1 + \ldots + P_k.$$

Take the trace of both sides: the $n \times n$ identity matrix I has trace n. Each P_i has trace its rank n_i, by Proposition 3.24, so

$$n = n_1 + \ldots + n_k.$$

Then squaring,

$$I = I^2 = \sum_i P_i^2 + \sum_{i<j} P_i P_j = \sum_i P_i + \sum_{i<j} P_i P_j.$$

Taking the trace,

$$n = \sum n_i + \sum_{i<j} \text{trace}(P_i P_j) = n + \sum_{i<j} \text{trace}(P_i P_j) :$$

$$\sum_{i<j} \text{trace}(P_i P_j) = 0.$$

Hence

$$
\begin{aligned}
\text{trace}(P_i P_j) &= \text{trace}(P_i^2 P_j^2) && \text{(since } P_i,\ P_j \text{ projections)} \\
&= \text{trace}((P_j P_i).(P_i P_j)) && \text{(trace}(AB) = \text{trace}(BA)) \\
&= \text{trace}((P_i P_j)^T.(P_i P_j)),
\end{aligned}
$$

since $(AB)^T = B^T A^T$ and P_i, P_j symmetric and where we have defined $A = P_i P_i P_j$, $B = P_j$. Hence we have that

$$\text{trace}(P_i P_j) \geq 0,$$

since for a matrix M

$$
\begin{aligned}
\text{trace}(M^T M) &= \sum_i (M^T M)_{ii} \\
&= \sum_i \sum_j (M^T)_{ij} (M)_{ji} \\
&= \sum_i \sum_j m_{ij}^2 \\
&\geq 0.
\end{aligned}
$$

So we have a sum of non-negative terms being zero. So each term must be zero. That is, the square of each element of $P_i P_j$ must be zero. So each element of $P_i P_j$ is zero, so matrix $P_i P_j$ is zero:

$$P_i P_j = 0 \qquad (i \neq j).$$

This is the condition that the *linear forms* $P_1 x, \ldots, P_k x$ be independent (Theorem 4.15 below). Since the $P_i x$ are independent, so are the $(P_i x)^T (P_i x) = x^T P_i^T P_i x$, that is, $x^T P_i x$ as P_i is symmetric and idempotent. That is, the *quadratic forms* $x^T P_1 x, \ldots, x^T P_k x$ are also independent.

We now have

$$x^T x = x^T P_1 x + \ldots + x^T P_k x.$$

The left is $\sigma^2 \chi^2(n)$; the ith term on the right is $\sigma^2 \chi^2(n_i)$.

We summarise our conclusions.

Theorem 3.27 (Chi-Square Decomposition Theorem)

If

$$I = P_1 + \ldots + P_k,$$

with each P_i a symmetric projection matrix with rank n_i, then

(i) the ranks sum:

$$n = n_1 + \ldots + n_k;$$

(ii) each quadratic form $Q_i := x^T P_i x$ is chi-square:

$$Q_i \sim \sigma^2 \chi^2(n_i);$$

(iii) the Q_i are mutually independent.

(iv)

$$P_i P_j = 0 \quad (i \neq j).$$

Property (iv) above is called *orthogonality* of the projections P_i; we study orthogonal projections in §3.6 below.

This fundamental result gives all the distribution theory that we shall use. In particular, since F-distributions are defined in terms of distributions of independent chi-squares, it explains why we constantly encounter F-statistics, and why all the tests of hypotheses that we encounter will be F-tests. This is so throughout the Linear Model – Multiple Regression, as here, Analysis of Variance, Analysis of Covariance and more advanced topics.

Note 3.28

The result above generalises beyond our context of projections. With the projections P_i replaced by symmetric matrices A_i of rank n_i with sum I, the corresponding result (Cochran's Theorem) is that (i), (ii) and (iii) are *equivalent*. The proof is harder (one needs to work with *quadratic* forms, where we were able to work with *linear* forms). For monograph treatments, see e.g. Rao (1973), §1c.1 and 3b.4 and Kendall and Stuart (1977), §15.16 – 15.21.

3.6 Orthogonal Projections and Pythagoras's Theorem

The least-squares estimators (LSEs) are the *fitted values*

$$\hat{y} = A\hat{\beta} = A(A^T A)^{-1} A^T y = AC^{-1} A^T y = Py,$$

with P the projection matrix (idempotent, symmetric) above. In the alternative notation, since P takes the data y into $\hat{y}$, P is called the *hat matrix*, and written H instead. Then

$$e := y - \hat{y} = y - Py = (I - P)y$$

('e for error') is the *residual vector*. Thus

$$y = A\beta + \epsilon = A\hat{\beta} + e = \hat{y} + e,$$

or in words,

data = true value + error = fitted value + residual.

Now

$$
\begin{aligned}
e^T \hat{y} &= y^T (I - P)^T Py \\
&= y^T (I - P)Py \quad (P \text{ symmetric}) \\
&= y^T (P - P^2)y \\
&= 0,
\end{aligned}
$$

as P is idempotent. This says that e, $\hat{y}$ are *orthogonal*. They are also both Gaussian (= multinormal, §4.3), as linear combinations of Gaussians are Gaussian (§4.3 again). For Gaussians, orthogonal = uncorrelated = independent (see § 4.3):

The residuals e and the fitted values $\hat{y}$ are independent
(see below for another proof). This result is of great practical importance, in the context of residual plots, to which we return later. It says that residual values e_i plotted against fitted values $\hat{y}_i$ should be *patternless*. If such a residual plot shows clear pattern on visual inspection, this suggests that our model may be wrong – see Chapter 7.

The data vector y is thus the hypotenuse of a right-angled triangle in n-dimensional space with other two sides the fitted values $\hat{y} = (I - P)y$ and the residual $e = Py$. The lengths of the vectors are thus related by Pythagoras's Theorem in n-space (Pythagoras of Croton, d. c497 BC):

$$\|y\|^2 = \|\hat{y}\|^2 + \|e\|^2.$$

In particular, $\|\hat{y}\|^2 \le \|y\|^2$:

$$\|\hat{P}y\|^2 \le \|y\|^2$$

for all y. We summarise this by saying that

$$\|P\| \leq 1$$

that is P has *norm* < 1, or P is *length-diminishing*. It is a projection from data-space (y-space) onto the vector subspace spanned by the least-squares estimates $\hat{\beta}$.

Similarly for $I - P$: as we have seen, it is also a projection, and by above, it too is length-diminishing. It projects from y-space onto the *orthogonal complement* of the vector subspace spanned by the LSEs.

For real vector spaces (as here), a projection P is *symmetric* ($P = P^T$) iff P is *length-diminishing* ($\|P\| \leq 1$) iff P is an *orthogonal*, or *perpendicular*, projection – the subspaces Im P and Ker P are orthogonal, or perpendicular, subspaces (see e.g. Halmos (1979), §75). Because our $P := AC^{-1}A^T$ ($C := A^T A$) is automatically symmetric and idempotent (a projection), this is the situation relevant to us.

Note 3.29

1. The use of the language, results and viewpoint of geometry – here in n dimensions – in statistics is ubiquitous in the Linear Model. It is very valuable, because it enables us to draw pictures and visualise, or 'see', results.

2. The situation in the Chi-Square Decomposition Theorem takes this further. There we have k (≥ 2) projections P_i summing to I, and satisfying the conditions

$$P_i P_j = 0 \qquad (i \neq j).$$

This says that the projections P_i are mutually *orthogonal*: if we perform two different projections, we reduce any vector to 0 (while if we perform the same projection *twice*, this is the same as doing it *once*). The P_i are *orthogonal projections*; they project onto *orthogonal subspaces*, L_i say, whose linear span is the whole space, L say:

$$L = L_1 \oplus \ldots \oplus L_k,$$

in the 'direct sum' notation $\oplus$ of Linear Algebra.

3. The case $k = 2$ is that treated above, with P, $I - P$ orthogonal projections and $L = L_1 \oplus L_2$, with $L_1 = \text{Im } P = \ker (I - P)$ and $L_2 = \text{Im } (I - P) = \ker P$.

Theorem 3.30

(i) $\hat{y} = Py \sim N(A\beta, \sigma^2 P)$.

(ii) $e := y - \hat{y} = (I - P)y \sim N(0, \sigma^2(I - P))$.

(iii) $e, \hat{y}$ are independent.

Proof

(i) $\hat{y}$ is a linear transformation of the Gaussian vector y, so is Gaussian. We saw earlier that the LSE $\hat{b}$ is unbiased for β, so $\hat{y} := A\hat{b}$ is unbiased for $A\beta$.

$$
\begin{aligned}
\mathrm{var}(\hat{y}) &= P\mathrm{var}(y)P^T \\
&= \sigma^2 PP^T && (\mathrm{var}(y) = \sigma^2 I) \\
&= \sigma^2 P^2 && (P \text{ symmetric}) \\
&= \sigma^2 P && (P \text{ idempotent}).
\end{aligned}
$$

(ii) Similarly e is Gaussian, mean 0 as $Ee = Ey - E\hat{y} = A\beta - A\beta = 0$.

$$
\begin{aligned}
\mathrm{var}(e) &= (I - P)\mathrm{var}(y)(I - P)^T \\
&= \sigma^2(I - P)(I - P^T) && (\mathrm{var}(y) = \sigma^2 I) \\
&= \sigma^2(I - P)^2 && (I - P \text{ symmetric}) \\
&= \sigma^2(I - P) && (I - P \text{ idempotent}).
\end{aligned}
$$

(iii)

$$
\begin{aligned}
\mathrm{cov}(\hat{y}, e) &= E\left[(\hat{y} - E\hat{y})^T(e - Ee)\right] \\
&= E\left[(\hat{y} - A\beta)^T e\right] && (E\hat{y} = A\beta,\ Ee = 0) \\
&= E\left[(Py - A\beta)^T(I - P)y\right] \\
&= E\left[\left(y^T P - \beta^T A^T\right)(y - Py)\right] \\
&= E[y^T Py] - E[y^T P^2 y] - \beta^T A^T Ey + \beta^T A^T A(A^T A)^{-1} A^T Ey \\
&= 0,
\end{aligned}
$$

using the idempotence of P. So $e, \hat{y}$ are uncorrelated, so independent (§4.3).

$\square$

Theorem 3.31

(i) $\hat{\beta} \sim N(\beta, \sigma^2 C^{-1})$.

(ii) $\hat{\beta}$ and SSE (or $\hat{\beta}$ and $\hat{\sigma}^2$) are independent.

Proof

(i) β is Gaussian; the mean and covariance were obtained in §3.3.

(ii) $\hat{\beta} - \beta = C^{-1}A^T(y - Ey) = C^{-1}A^T(y - A\beta)$ and $SSE = (y - A\beta)^T(I - P)(y - A\beta)$, above. Now since $(I - P)^2 = I - P$,

$$((I-P)(y-A\beta))^T((I-P)(y-A\beta)) = (y-A\beta)^T(I-P)(y-A\beta) = SSE,$$

so it suffices to prove that $C^{-1}A^T(y - A\beta)$ and $(I - P)(y - A\beta)$ are independent. Since the covariance matrix of $y - A\beta$ is $\sigma^2 I$, and

$$C^{-1}A^T.(I - P) = C^{-1}A^T - C^{-1}A^T.AC^{-1}A^T = 0,$$

this follows from the criterion for independence of linear forms in §4.3 below.

$\square$

This yields another proof of:

Corollary 3.32

SSR and SSE are independent.

Proof

$SSR := (\hat{\beta} - \beta)^T C(\hat{\beta} - \beta)$ is a function of $\hat{\beta}$, so this follows from (ii) above.

$\square$

Finally, Theorem 3.31 also gives, when combined with Theorem 3.26, a method for calculating one-dimensional confidence intervals for the individual elements of β. We have

Corollary 3.33

Let β_i denote the ith element of β and C_{ii}^{-1} the ith diagonal element of C^{-1}. We have

$$\frac{\beta_i - \hat{\beta}_i}{\hat{\sigma}\sqrt{C_{ii}^{-1}}} \sim t_{n-p}.$$

Proof

From Theorem 3.26 we have that

$$\hat{\sigma}^2 \sim \frac{\sigma^2 \chi^2}{n-p}.$$

Further $\beta - \hat{\beta}$ is $N(0, \sigma^2 C^{-1})$ and is independent of $\hat{\sigma}$. The stochastic representation

$$\frac{\beta - \hat{\beta}}{\hat{\sigma}} = \frac{N(0, C^{-1})}{\sqrt{\chi^2_{n-p}/(n-p)}},$$

where $N(0, C^{-1})$ and χ^2_{n-p} denote independent random variables with the multivariate normal and univariate χ^2_{n-p} distributions, can be seen to lead to a multivariate Student t distribution (Exercise 3.10). The full result follows by considering the properties of the univariate marginals of this distribution and is left to the reader (Exercise 3.10). □

3.7 Worked examples

We turn below to various examples. The first thing to do is to identify the design matrix A, and then find the various matrices – particularly the projection matrix P – associated with it. The first example is small enough to do by hand, but large enough to be non-trivial and to illustrate the procedure.

Example 3.34

Two items A and B are weighed on a balance, first separately and then together, to yield observations y_1, y_2, y_3.

1. *Find the LSEs of the true weights β_A, β_B.*

   We have

   $$\begin{aligned} y_1 &= \beta_A + \epsilon_1, \\ y_2 &= \beta_B + \epsilon_2, \\ y_1 + y_2 &= \beta_A + \beta_B + \epsilon_3, \end{aligned}$$

with errors ϵ_i iid $N(0, \sigma^2)$. So

$$Ey = \begin{pmatrix} 1 & 0 \\ 0 & 1 \\ 1 & 1 \end{pmatrix} \cdot \begin{pmatrix} \beta_A \\ \beta_B \end{pmatrix}.$$

The design matrix is thus

$$A = \begin{pmatrix} 1 & 0 \\ 0 & 1 \\ 1 & 1 \end{pmatrix}.$$

So

$$C = A^T A = \begin{pmatrix} 1 & 0 & 1 \\ 0 & 1 & 1 \end{pmatrix} \cdot \begin{pmatrix} 1 & 0 \\ 0 & 1 \\ 1 & 1 \end{pmatrix} = \begin{pmatrix} 2 & 1 \\ 1 & 2 \end{pmatrix}.$$

So $|C| = 3$, and

$$C^{-1} = \frac{1}{3} \begin{pmatrix} 2 & -1 \\ -1 & 2 \end{pmatrix},$$

$$A^T y = \begin{pmatrix} 1 & 0 & 1 \\ 0 & 1 & 1 \end{pmatrix} \cdot \begin{pmatrix} y_1 \\ y_2 \\ y_3 \end{pmatrix} = \begin{pmatrix} y_1 + y_3 \\ y_2 + y_3 \end{pmatrix},$$

$$\begin{aligned} \hat{\beta} &= C^{-1} A^T y = \frac{1}{3} \begin{pmatrix} 2 & -1 \\ -1 & 2 \end{pmatrix} \begin{pmatrix} y_1 + y_3 \\ y_2 + y_3 \end{pmatrix}, \\ &= \frac{1}{3} \begin{pmatrix} 2y_1 - y_2 + y_3 \\ -y_1 + 2y_2 + y_3 \end{pmatrix}. \end{aligned}$$

The first and second components of this 2-vector are the required LSEs of β_A and β_B.

2. *Find the covariance matrix of the LSE.*

   This is

$$\mathrm{var}(\hat{\beta}) = \sigma^2 C^{-1} = \frac{\sigma^2}{3} \begin{pmatrix} 2 & -1 \\ -1 & 2 \end{pmatrix}.$$

3. *Find SSE and estimate σ^2.*

$$\begin{aligned} P = A.C^{-1}A^T &= \begin{pmatrix} 1 & 0 \\ 0 & 1 \\ 1 & 1 \end{pmatrix} \cdot \frac{1}{3} \begin{pmatrix} 2 & -1 \\ -1 & 2 \end{pmatrix} \begin{pmatrix} 1 & 0 & 1 \\ 0 & 1 & 1 \end{pmatrix} \\ &= \frac{1}{3} \begin{pmatrix} 2 & -1 & 1 \\ -1 & 2 & 1 \\ 1 & 1 & 2 \end{pmatrix}, \end{aligned}$$

$$I - P = \frac{1}{3} \begin{pmatrix} 1 & 1 & -1 \\ 1 & 1 & -1 \\ -1 & -1 & 1 \end{pmatrix}.$$

So

$$
\begin{aligned}
SSE &= y^T(I - P)y, \\
&= \frac{1}{3} \begin{pmatrix} y_1 & y_2 & y_3 \end{pmatrix} \begin{pmatrix} 1 & 1 & -1 \\ 1 & 1 & -1 \\ -1 & -1 & 1 \end{pmatrix} \cdot \begin{pmatrix} y_1 \\ y_2 \\ y_3 \end{pmatrix}, \\
&= \frac{1}{3}(y_1 + y_2 - y_3)^2.
\end{aligned}
$$

Since $n = 3$, $p = 2$, $n - p = 1$ here, this is also $\hat{\sigma}^2$:

$$\hat{\sigma}^2 = \frac{1}{3}(y_1 + y_2 - y_3)^2.$$

Example 3.35 (Simple linear regression via multiple linear regression)

We illustrate how multiple regression generalises the simple linear regression model of Chapter 1. In the notation of Lemma 3.3 and since $y_i = \alpha + \beta x_i + \epsilon_i$ $(i = 1, \ldots, n)$

$$y = A\beta + \epsilon,$$

where $A = \begin{pmatrix} 1 & x_1 \\ \vdots & \vdots \\ 1 & x_n \end{pmatrix}$ and $\beta = \begin{pmatrix} \alpha \\ \beta \end{pmatrix}$. We see that

$$C = A^T A = \begin{pmatrix} 1 & \cdots & 1 \\ x_1 & \cdots & x_n \end{pmatrix} \begin{pmatrix} 1 & x_1 \\ \vdots & \vdots \\ 1 & x_n \end{pmatrix} = \begin{pmatrix} n & \sum x \\ \sum x & \sum x^2 \end{pmatrix}.$$

Further, we can deduce from the fact that $|C| = n\sum x^2 - (\sum x)^2$ that $|C| > 0$ by the Cauchy–Schwarz inequality. Hence C is invertible with

$$C^{-1} = \frac{1}{n\sum x^2 - (\sum x)^2} \begin{pmatrix} \sum x^2 & -\sum x \\ -\sum x & n \end{pmatrix}.$$

It follows that

$$A^T y = \begin{pmatrix} 1 & \cdots & 1 \\ x_1 & \cdots & x_n \end{pmatrix} \begin{pmatrix} y_1 \\ \vdots \\ y_n \end{pmatrix} = \begin{pmatrix} \sum y \\ \sum xy \end{pmatrix}.$$

The solution for $\hat{\beta}$ becomes

$$
\begin{aligned}
\hat{\beta} &= C^{-1}A^T y = \frac{1}{n\sum x^2 - (\sum x)^2} \left(\begin{array}{cc} \sum x^2 & -\sum x \\ -\sum x & n \end{array} \right) \left(\begin{array}{c} \sum y \\ \sum xy \end{array} \right), \\
&= \frac{1}{\sum x^2 - n\overline{x}^2} \left(\begin{array}{c} \sum x^2 \overline{y} - \overline{x}\sum xy \\ \sum xy - n\overline{x}\overline{y} \end{array} \right), \\
&= \frac{1}{\overline{(x^2)} - (\overline{x})^2} \left(\begin{array}{c} \overline{(x^2)}\overline{y} - \overline{x}\overline{(xy)} \\ \overline{(xy)} - \overline{x}\overline{y} \end{array} \right),
\end{aligned}
$$

dividing top and bottom by n. The second coordinate gives

$$
\hat{\beta} = s_{xy}/s_{xx},
$$

as before. Adding and subtracting $\overline{y}(\overline{x})^2$, the first coordinate gives

$$
\begin{aligned}
\hat{\alpha} &= \left(\overline{y}\left[\overline{(x^2)} - (\overline{x})^2 \right] - \overline{x}\left[\overline{(xy)} - \overline{x}\overline{y} \right] \right)/s_{xx} = (\overline{y} - \overline{x}s_{xy})/s_{xx} \\
&= \overline{y} - \overline{x}s_{xy}/s_{xx} = \overline{y} - \hat{\beta}\overline{x},
\end{aligned}
$$

as before.

We illustrate multiple regression models with two 'more statistical' examples.

Example 3.36 (Athletics times: snapshot data)

While athletic performance is much more variable within sexes than between sexes, men are nevertheless faster on average than women. This *gender effect* is caused by basic physiology, such as pelvis design. As regular competitors in distance races will know, there is also a *club effect*: club members are on average faster than non-club members (mainly because of the benefits of training with club mates, but there may also be selection bias, in that the better athletes are more likely to join a club). Age is also important. There are three phases in an athlete's life: development, plateau, and eventual decline with age. For distance running, as remarked earlier, there is for this *age effect* a well-known runner's Rule of Thumb: for every year into the decline phase, one can expect to lose a minute a year on the marathon through age alone (and pro rata for shorter distances).

One may seek to use regression to do two things:
(i) confirm and quantify these gender, club and age effects;
(ii) assess the proportion of variability in athletes' performance accounted for by knowledge of sex, club status and age.

We take as the basis of our discussion the analysis in Bingham and Rashid (2008). This study uses six years of data (2002-2007) for the Berkhamsted Half

Marathon and regresses time in minutes against age and indicator variables representing gender (0=Male) and club status (1=Member) see (Exercise 1.6). Summary results for analysis of this data are given in Table 3.1.

Year	Intercept	Club	Gender	Age	R^2
2002	75.435	-7.974	15.194	0.504	0.231
2003	74.692	-9.781	14.649	0.534	0.200
2004	75.219	-9.599	17.362	0.406	0.274
2005	74.401	-10.638	16.788	0.474	0.262
2006	86.283	-9.762	13.002	0.312	0.198
2007	91.902	-11.401	14.035	0.192	0.177

Table 3.1 Regression results for Example 3.36

It is clear that most of the variability observed is variability between athletes, caused by innumerable factors, principally innate ability and training, rather than age. Nonetheless a non-trivial proportion of the observed variability ($\approx 22\%$) can be explained by knowledge of club status, age and gender. The estimates in Table 3.1 lead to sensible conclusions and suggest that club members tend to be faster than non-club members (by 9 to 10 minutes), men tend to be faster than women (by 13 to 15 minutes) and increased age leads to slower times.

Example 3.37 (Athletics times: One athlete)

One way to focus on the age effect is to reduce the data to one athlete over time, where ageing can be studied directly and there is no between-athlete variability. For convenience, we use the data set in Table 1.1. We consider a power-law model

$$t = cd^{b_1} a^{b_2},$$

with t, d, a, representing time, distance and age respectively, b_1, b_2 the exponents and c a parameter measuring the individual athlete's quality or speed. So

$$\frac{\partial t}{\partial a} = b_2 \frac{t}{a}.$$

This may be handled via a linear model by setting

$$\log t = \log c + b_1 \log d + b_2 \log a + \epsilon.$$

Estimates for this model are summarised in Table 3.2.

	Value	Std. Error	t value
Intercept	0.547	0.214	2.551
log(age)	0.332	0.051	6.471
log(distance)	1.017	0.015	66.997

Table 3.2 Regression results for Example 3.37

Here, t is about 90, a is about 60. So t/a is about $3/2$, but from the model output in Table 3.2, b_2 is about $1/3$, so $\partial t/\partial a$ is about $1/2$. Thus (at least for athletes of about this age and quality) one can expect to lose half a minute on the half-marathon per year through ageing alone, or a minute a year in the marathon – in good agreement with the Rule of Thumb.

EXERCISES

3.1. An athlete runs 800m in a time trial. Three time keepers time him, for the first 400m (y_1), the second 400m (y_2), and for the whole 800m (y_3). Estimate

   a) his true times for each of the two laps,

   b) the accuracy of the time keepers.

   This is the balance example (Example 3.34) in a different guise.

3.2. A castle, which stands on a flat base, has four towers. Six independent readings were taken – five differences in height and the height of the shortest tower G – measured in metres. The data were $D - G = 12.29$, $F - D = 24.46$, $E - D = 20.48$, $F - E = 3.59$, $F - G = 36.32$ and $D = 46.81$. Calculate
   (i) The matrices A, $C = A^T A$, $P = A C^{-1} A^T$,
   (ii) The least squares estimates of the true tower heights, together with an unbiased estimate of the standard deviation σ.

3.3. *Adjusted R^2 statistic.*

$$R_a^2 = 1 - (1 - R^2)\left(\frac{n-1}{n-p}\right).$$

   (i) Using the definition, show that if model 1 has p parameters and model 2 has $p + 1$ parameters, the criterion for rejecting model 1 in favour of model 2 becomes

$$R_2^2 > 1 - \frac{(1 - R_1^2)(n - 1 - p)}{(n - p)}.$$

(ii) What does this condition become when model 2 has j additional parameters?

3.4. *Artificial data set.* A simulated data set linking a response variable Y to explanatory variables X and Z is shown in Table 3.3.
(i) Plot Y against Z. Does a quadratic term in Z appear reasonable?
(ii) Fit the model $Y = a + bX + cZ + dZ^2$ and comment on the results. (A more in-depth approach to finding a suitable model will require the methods of Chapter 7.)

3.5. *Cherry tree data.* The volumes of 31 cherry trees were recorded along with their girths and heights. The data are shown in Table 3.4. It is desired to predict volume v based on measurements of girth g and height h.
(i) Does it seem necessary to include quadratic terms in g and h in the model? Consider both t-statistics and exploratory plots.
(ii) By thinking of trees as roughly cylindrical, suggest a possible model for v. Fit this model and compare with the models in (i).

3.6. *Matrix calculus.* From first principles derive the relations

$$\text{(i)} \quad \frac{\partial a^T x}{\partial x} = a^T, \quad \frac{\partial (Ax)}{\partial x} = A;$$

$$\text{(ii)} \quad \frac{\partial (x^T Ax)}{\partial x} = x^T (A^T + A).$$

3.7. *Derivation of normal equations/ordinary least squares solution via matrix calculus.* Show that this can be achieved by minimising the sum of squares

$$SS := (y - A\beta)^T (y - A\beta)$$

as a function of β. You may use Exercise 3.6 as appropriate.

3.8. *Gram–Schmidt process.* Use the Gram–Schmidt process to produce an orthonormal basis of the linear subspace spanned by $a_1 = (-2, -1, -2, 0)^T$, $a_2 = (2, 2, 2, 1)^T$, $a_3 = (-2, -2, -1, -1)^T$.

3.9. *QR decomposition.* Using the QR decomposition provide an alternative derivation of the estimates of a and b in the simple linear regression model in Chapter 1.

Y	X	Z	Y	X	Z	Y	X	Z
15.42	5.0	3.05	82.75	8.0	3.91	15.42	11.5	2.94
47.33	5.0	5.77	37.03	8.5	3.00	49.94	11.5	3.73
34.36	5.0	4.29	43.38	8.5	3.28	68.40	11.5	4.36
44.44	5.0	3.99	23.92	8.5	2.31	20.03	12.0	2.94
11.04	5.0	2.66	24.50	8.5	2.64	72.20	12.0	4.75
9.67	5.5	1.46	16.53	8.5	2.47	14.85	12.0	1.69
39.29	5.5	3.15	18.92	9.0	2.74	115.36	12.0	4.81
13.14	5.5	2.83	22.57	9.0	2.72	21.09	12.0	2.72
30.33	5.5	3.01	0.30	9.0	3.41	51.02	12.5	4.23
14.56	5.5	2.63	18.00	9.0	2.94	22.40	12.5	3.10
11.22	6.0	2.03	31.88	9.0	3.54	24.11	12.5	3.24
15.58	6.0	2.63	37.09	9.5	3.20	21.45	12.5	2.59
11.59	6.0	2.09	20.90	9.5	2.70	48.62	12.5	3.88
10.53	6.0	1.49	73.03	9.5	4.03	21.21	13.0	2.42
17.09	6.0	2.70	32.38	9.5	3.04	22.82	13.0	3.31
64.46	6.5	3.88	28.98	9.5	3.15	24.34	13.0	2.87
66.16	6.5	4.54	25.34	10.0	2.78	15.02	13.0	2.44
21.94	6.5	2.74	19.18	10.0	2.62	12.92	13.0	1.93
32.46	6.5	3.78	30.38	10.0	3.62	22.43	13.5	2.30
28.25	6.5	3.87	43.87	10.0	3.69	56.61	13.5	3.21
26.68	7.0	3.39	12.77	10.0	2.31	16.54	13.5	1.86
19.99	7.0	3.03	40.32	10.5	3.53	36.38	13.5	3.25
81.67	7.0	3.78	33.31	10.5	3.72	20.95	13.5	2.17
46.84	7.0	3.31	18.11	10.5	2.24	44.77	14.0	3.90
12.42	7.0	1.90	26.25	10.5	2.47	18.25	14.0	2.24
22.98	7.5	2.50	58.39	10.5	4.28	33.23	14.0	3.30
44.86	7.5	4.60	4.65	11.0	2.99	41.20	14.0	3.60
33.33	7.5	3.51	13.45	11.0	1.97	26.55	14.0	2.17
49.80	7.5	3.91	36.55	11.0	4.10	13.38	14.5	3.16
16.75	7.5	2.24	14.04	11.0	1.49	28.82	14.5	2.30
18.43	8.0	2.18	31.63	11.0	3.20	28.06	14.5	2.99
46.13	8.0	3.42	54.46	11.5	3.56	17.57	14.5	2.50
23.97	8.0	2.73	38.06	11.5	3.37	18.71	14.5	2.33
38.75	8.0	3.49						

Table 3.3 Data for Exercise 3.4

3.10. Analogously to Exercise 1.11, and using the same notation, one can define an r-dimensional multivariate t-distribution by

1. Generate u from f_Y
2. Generate x from $N(0, u\Delta)$ for some 'correlation' matrix Δ.

Volume	0.7458, 0.7458, 0.7386, 1.1875, 1.3613, 1.4265, 1.1296, 1.3179, 1.6365, 1.4410, 1.7524, 1.5206, 1.5496, 1.5424, 1.3831, 1.6075, 2.4475, 1.9841, 1.8610, 1.8030, 2.4982, 2.2954, 2.6285, 2.7734, 3.0847, 4.0116, 4.0333, 4.2216, 3.7292, 3.6930, 5.5757
Girth	66.23, 68.62, 70.22, 83.79, 85.38, 86.18, 87.78, 87.78, 88.57, 89.37, 90.17, 90.97, 90.97, 93.36, 95.76, 102.94, 102.94, 106.13, 109.32, 110.12, 111.71, 113.31, 115.70, 127.67, 130.07, 138.05, 139.64, 142.84, 143.63,143.63, 164.38
Height	21.0, 19.5, 18.9, 21.6, 24.3, 24.9, 19.8, 22.5, 24.0, 22.5, 23.7, 22.8, 22.8, 20.7, 22.5, 22.2, 25.5, 25.8, 21.3, 19.2, 23.4, 24.0, 22.2, 21.6, 23.1, 24.3, 24.6, 24.0, 24.0, 24.0, 26.1

Table 3.4 Data for Exercise 3.5.

(i) Using the construction in Exercise 1.11 show that the univariate marginals of this distribution satisfy

$$\frac{x_i}{\sqrt{\Delta_{ii}}} \sim t_r.$$

The above argument can be used to show that

$$\frac{\beta_i - \hat{\beta}_i}{\sqrt{(X^T X)_{ii}^{-1}}} \sim t_{n-p}.$$

Derive the estimated standard errors (e.s.e.) for parameters in the following models:
(ii) The simple linear regression model.
(iii) The bivariate regression model.

3.11. *Prediction intervals for a future observation.* In the linear regression model, the underlying mean corresponding to an observation $X = X_0^T$ is $E[Y] = X_0^T \beta$. In practice, we estimate $E[Y] = X_0^T \hat{\beta}$ with associated variance $\sigma^2(X_0^T(X^T X)^{-1}X_0)$. $100(1 - \alpha)\%$ confidence intervals for the underlying mean can be constructed to give

$$X_0^T \hat{\beta} \pm t_{n-p}(1 - \alpha/2)\hat{\sigma}\sqrt{X_0^T(X^T X)^{-1}X_0}.$$

Suppose a future observation is to be taken as $X = X_0^T$. Amend the above procedure to produce an appropriate confidence interval.

3.12. *Commutativity of matrix trace.* Show that

$$\text{trace}(AB) = \text{trace}(BA).$$

4

Further Multilinear Regression

4.1 Polynomial Regression

For one regressor x, simple linear regression is fine for fitting straight-line trends. But what about more general trends – quadratic trends, for example? (E.g. height against time for a body falling under gravity is quadratic.) Or cubic trends? (E.g.: the van der Waals equation of state in physical chemistry.) Or quartic? – etc.

We can use the successive powers $x^0 = 1$, x, $x^2, \ldots$ as regressors, so that a polynomial is a special case of multilinear regression.

It is important to note that, although a polynomial of degree higher than one is *non*-linear in the variable x, it is *linear* in the coefficients, which serve here as the parameters. Indeed, we encountered an instance of this in §1.2, in connection with the work of Legendre and Gauss – fitting elliptical orbits by least squares.

Recall that in a regression model, we are seeking to decompose our data into a systematic component and a random component. We will only go beyond the linear regression of Chapter 1 if a linear fit is poor. We then seek to improve the fit by adding more terms. However, it is very important to notice that one should not go too far here. Let us assume for the moment that all the x-values $x_1, \ldots, x_n$ are *distinct*. Then we can achieve an *exact* fit with a polynomial of degree $n - 1$, which contains n coefficients. Of course, if there are coincident x-values an exact fit is clearly impossible, as the corresponding y-value can only fit *one* of the x-values. The fact that with distinct x-values an exact fit is

N.H. Bingham and J.M. Fry, *Regression: Linear Models in Statistics*,
Springer Undergraduate Mathematics Series, DOI 10.1007/978-1-84882-969-5_4,
© Springer-Verlag London Limited 2010

indeed possible is a result from the important subject of Interpolation, a branch of Numerical Analysis. But fitting exactly with an $(n-1)$-degree polynomial would be very foolish. For, n is large, so $n-1$ is also large, and *polynomials of large degree have very bad numerical properties*. One way to see this is to examine the tendency of characteristic polynomials – used to find eigenvalues – to change their roots dramatically with only small changes in their coefficients (such as are inevitably caused by numerical rounding error). For a monograph treatment of this subject, we refer to the classic Wilkinson (1965).

Note 4.1

One can combine the good numerical properties of polynomials of low degree with the many degrees of freedom of polynomials of high degree by using *splines*. These are separate polynomials on separate ranges of the x-variable, spliced together at the points separating the sub-intervals – the *knots* – so as to be continuous, and have $k-1$ derivatives continuous, where k is the degree of polynomial in use; see §9.2. Thus for $k=1$ a linear spline is a piecewise-continuous linear function (a 'broken line'); for a quadratic spline we have one derivative continuous also, and for the *cubic splines*, in very common use, we have two derivatives continuous. Splines are extensively used in non-linear regression and smoothing in Statistics, in Numerical Analysis, and elsewhere. We will return to splines later in §9.2 on non–parametric regression. In fact smoothing splines have now largely replaced polynomials in regression in practice, but we need to learn to walk before we learn to run.

Recall that in regression we have

$$\mathbf{data} = \mathbf{signal} + \mathbf{noise} = \mathbf{trend} + \mathbf{error}.$$

Our job is to *reveal* the trend by *removing* the error – or as much of it as we can. In the context of polynomial regression, we are caught between two opposing dangers. If we take the degree of the polynomial too low – fit a linear trend through data which comes from a perturbed quadratic, say – we *distort* the trend. If on the other hand we take the degree too high, we leave in too much error, and instead *obscure* the trend. This is called *over-interpretation*, or *over-fitting*. It has the effect of treating the data – which, being obtained by sampling, inevitably contains random sampling error – with 'too much respect'. Instead, we should exploit our main advantage – that n is large, and so the Law of Large Numbers, the tendency of independent errors to cancel, works on our side.

The question raised by all this is how to choose the degree $p-1$ (p parameters). The formal question of testing the hypothesis that the leading term is actually needed we defer to Chapter 6. An informal treatment suffices for our present purposes. First, by EDA as usual, plot the data, inspect visually, and decide what is the highest order of polynomial we would be prepared to consider (four or five – corresponding to five or six parameters – would be as high as one would normally be prepared to go). Then use a statistical package to perform the regression, and inspect the printout for significance of coefficients. A good standard package – Minitab® or S-Plus/R®, for example – will print out, by the side of each coefficient estimate, a probability that it could be as big as this by chance alone. A probability that vanishes to several places of decimals indicates a term that is highly significant, and that we clearly need. A probability of 0.3, say, indicates a term that could easily have arisen by chance alone, and this suggests that our model could do better without it – and so, would be better without it (see §4.1.1 below).

Example 4.2 (Polynomial regression)

The data in Table 4.1 link the yield X to the percentage protein content Y of an agricultural experiment. The layout below clearly shows that a nonlinear model in X can be handled by a model that remains linear in the parameters.

X	5, 8, 10, 11, 14, 16, 17, 17, 18, 20, 22, 24, 26, 30, 32, 34, 36, 38, 43
X^2	25, 64, 100, 121, 196, 256, 289, 289, 400, 484, 576, 676, 900, 1024, 1156, 1296, 1444, 1849
X^3	125, 512, 1000, 1331, 2744, 4096, 4913, 4913, 5832, 8000, 10648, 13824, 17576, 27000, 32768, 39304, 46656, 54872, 79507
Y	16.2, 14.2, 14.6, 18.3, 13.2, 13.0, 13.0, 13.4, 10.6, 12.8, 12.6, 11.6, 11.0, 9.8, 10.4, 10.9, 12.2, 9.8, 10.7

Table 4.1 Data for Example 4.2

A plot of the data is shown in Figure 4.1 and there is at least some indication of a nonlinear relationship between X and Y.

Simple linear regression model: $Y = a + bX$. The t-test gives a p-value of 0.000, indicating that the X term is needed in the model. The R^2 value is a reasonably high 0.61.

Quadratic regression model: $Y = a + bX + cX^2$. The R^2 value increases to 0.701. The univariate t-test gives a p-value of 0.004 that $b = 0$, and $p = 0.043$ that $c = 0$. Thus it appears that both quadratic and linear terms are needed in the model.

Cubic regression model: $Y = a + bX + cX^2 + dX^3$. The univariate t-test gives a p-value of 0.733 that $d = 0$. The R^2 value is 0.703, only a marginal improvement on the quadratic model.

In conclusion we have some suggestion of a nonlinear relationship between X and Y, and a skeleton analysis suggests a quadratic model might be appropriate.

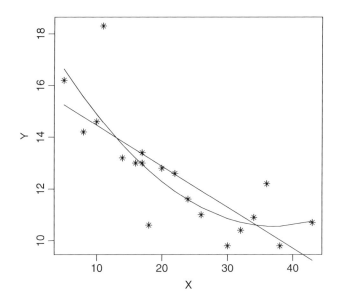

Figure 4.1 Plot of Y against X together with lines of best fit for both linear and quadratic models.

4.1.1 The Principle of Parsimony

The great Albert Einstein (1879–1955) had a famous dictum: *Physics should be made as simple as possible, but not simpler*. Einstein's Dictum applies of course with as much force to Mathematics, to Statistics, or to any other branch of science. It is prefigured by *Occam's Razor*: *Entities are not to be multiplied without necessity* (William of Ockham (d. c1349), quoted from Jeffreys (1983), 342). In brief: if we *can* do without something, we *should* do without it.

The modern form of this admirable precept is known in Statistics as the *Principle of Parsimony*. It encourages us to use simple models for preference, thus gaining in both clarity and in protection against the danger of over-interpretation. It also suggests the idea of *penalising models for complexity*, which we formalise in the Akaike Information Criterion (AIC) of §5.2.1 below.

4.1.2 Orthogonal polynomials

Suppose we begin with a linear regression, but then decide to change to a quadratic – or wish to increase from a quadratic to a cubic, etc. We have to begin again from scratch. It would be preferable to have a situation in which adding an extra term merely *refined* the model – by increasing its order of accuracy – rather than *changing* it completely. (In pre-computer days this was even more important – calculation had to be done by hand and was time-consuming. This is less important nowadays, with computer packages taking care of the calculation, but is still important conceptually.) We can do this by using, not the *powers* $x^0 = 1, x, x^2, x^3, \ldots$ in succession, but a system of *orthogonal polynomials* (OPs), the kth of which has degree k. These may be constructed by the process of *Gram–Schmidt orthogonalisation* (used in Linear Algebra to turn a spanning set into an orthogonal basis). For details, see e.g. Plackett (1960), Ch. 6. This idea is also developed in §5.1.1 below on orthogonal parameters.

Note 4.3

The reader has probably already encountered orthogonal polynomials of special kinds in other contexts. The classical cases of Legendre, Hermite and Tchebycheff polynomials, for example, are commonly used in Applied Mathematics. These are all examples of *continuous* orthogonal polynomials (where the orthogonality relation involves integrating), whereas in our situation we have *discrete* orthogonal polynomials (where the orthogonality relation involves summation). The two cases may be handled together. Orthogonal polynomials are very useful and have an extensive and interesting theory; the classic monograph is Szegö (1959).

4.1.3 Packages

In Minitab®, one declares the powers one wishes to use as regressors. For a detailed worked example of polynomial regression in Minitab®, see e.g. Joiner and Ryan (2000) §10.6 and §10.7. In S-Plus, one uses `lm` for regression as usual, and `poly`, specifying the degree. Thus

$$\mathtt{lm(y \sim poly(x, 3))}$$

fits a polynomial in x of degree 3 to the data y. The default option in S-Plus uses orthogonal polynomials. If we wish to work with ordinary powers, one uses `poly.transform`, thus:

$$\mathtt{xylm < -lm(y \sim poly(x, 3))}$$

$$\mathtt{poly.transform(poly(x, 3), coef(xylm))}$$

A worked example is included in the online S-Plus Help facility, p. 275 (for plot) and 278–9 (for S-Plus output and discussion). Here y is NO (nitric oxide), x is E (ethanol); the plot looks quadratic, but with 'turned-up ends', and analysis indicates using a quartic fit.

4.2 Analysis of Variance

We illustrate the richness of the class of linear models by illustrating how the Analysis of Variance of Chapter 2 can be reformulated as a regression model within the linear models framework of Chapter 3. Suppose that we have a one-way ANOVA model with k groups:

$$y_{ij} = \mu + \alpha_i + \epsilon_{ij} \quad (j = 1, \ldots, k).$$

If we define $\mathbf{y}_i$ to be the vector of observations y_{ij} from treatment i, we might imagine that this might be formulated as a regression model using

$$\mathbf{y} = \begin{pmatrix} \mathbf{y}_1 \\ \mathbf{y}_2 \\ \ldots \\ \mathbf{y}_k \end{pmatrix}, \quad \mathbf{A} = \begin{pmatrix} 1_{n_1} & 1_{n_1} & 0_{n_1} & \ldots & 0_{n_1} \\ 1_{n_2} & 0_{n_2} & 1_{n_2} & \ldots & 0_{n_2} \\ \ldots & \ldots & \ldots & \ldots & \ldots \\ 1_{n_k} & 0_{n_k} & 0_{n_k} & 0_{n_k} & 1_{n_k} \end{pmatrix}, \quad \beta = \begin{pmatrix} \mu \\ \alpha_1 \\ \ldots \\ \alpha_k \end{pmatrix},$$

where 1_{n_i} is a n_i vector of 1s corresponding to the n_i observations in treatment group i. Note, however, that under this formulation $\mathbf{A}^T \mathbf{A}$ is not invertible. (If we let a be the n-dimensional column vector $a = (-1, 1, \ldots, 1)^T$ then $\mathbf{A}a = \mathbf{0}$. If $C = A^T A$, $a^T C a = 0$ and C is not positive definite.) There are a number of ways in which this model can be reparametrised to remove the linear dependency in the columns of the X matrix (see Exercise 4.2). One way of doing this is to set $\mu = 0$. Under this formulation we have that

$$
A = \begin{pmatrix} 1_{n_1} & 0_{n_1} & \dots & 0_{n_1} \\ 0_{n_2} & 1_{n_2} & \dots & 0_{n_2} \\ \dots & \dots & \dots & \dots \\ 0_{n_k} & 0_{n_k} & \dots & 1_{n_k} \end{pmatrix}, \quad \beta = \begin{pmatrix} \alpha_1 \\ \alpha_2 \\ \dots \\ \alpha_k \end{pmatrix}.
$$

We proceed to show that this formulation returns exactly the same results for Analysis of Variance as in Chapter 2. From Theorem 3.31 it follows that $SSE = y^T(I - AC^{-1}A^T)y$. We see that

$$
C = \begin{pmatrix} n_1 & 0 & \dots & 0 \\ 0 & n_2 & \dots & 0 \\ 0 & 0 & \dots & 0 \\ 0 & 0 & \dots & n_k \end{pmatrix}, \quad A^T y = \begin{pmatrix} n_1 \bar{y}_1 \\ \dots \\ n_k \bar{y}_k \end{pmatrix}.
$$

From the above it follows that C^{-1} is the diagonal matrix with elements $1/n_i$ and that the fitted values are given by $\hat{\beta} = C^{-1}A^T y = (\bar{y}_1, \dots, \bar{y}_k)^T$, which are intuitive and correspond to the fitted values given by ANOVA. The appropriate F-test to test for the difference in the mean values can be recovered by the methods of Chapter 6 (Exercise 6.8).

Alternatively, one could have proceeded by retaining μ and setting one of the α_i, α_1 say, equal to zero. See Exercise 4.2. Both approaches give a general technique through which categorical or discrete variables can be incorporated into these and related models. The comment is made in Draper and Smith (1998) Ch. 23 that in practice it is usually expedient to use specialist software and methods to fit Analysis of Variance models. However, the comparison with linear models reinforces that Analysis of Variance models are subject to exactly the same kind of model checks and scrutiny as other linear models; see Chapter 7. Further, this approach also motivates an extension – Analysis of Covariance – which includes discrete and continuous variables in the same model (see Chapter 5).

4.3 The Multivariate Normal Distribution

With one regressor, we used the bivariate normal distribution as in Chapter 1. Similarly for two regressors, we used – implicitly – the trivariate normal. With any number of regressors, as here, we need a general *multivariate normal*, or *'multinormal'*, or *Gaussian* distribution in n dimensions. We must expect that in n dimensions, to handle a random n-vector $\mathbf{X} = (X_1, \cdots, X_n)^T$, we will need

(i) a *mean vector* $\mu = (\mu_1, \ldots, \mu_n)^T$ with $\mu_i = EX_i$, $\mu = E\mathbf{X}$,

(ii) a *covariance matrix* $\Sigma = (\sigma_{ij})$, with $\sigma_{ij} = \mathrm{cov}(X_i, X_j)$, $\Sigma = \mathrm{cov}\mathbf{X}$.

First, note how mean vectors and covariance matrices transform under linear changes of variable:

Proposition 4.4

If $\mathbf{Y} = A\mathbf{X} + \mathbf{b}$, with $\mathbf{Y}, \mathbf{b}$ m-vectors, A an $m \times n$ matrix and $\mathbf{X}$ an n-vector,

(i) the mean vectors are related by $E\mathbf{Y} = AE\mathbf{X} + \mathbf{b} = A\mu + \mathbf{b}$,

(ii) the covariance matrices are related by $\Sigma_Y = A\Sigma_X A^T$.

Proof

(i) This is just linearity of the expectation operator E: $Y_i = \sum_j a_{ij} X_j + b_i$, so

$$EY_i = \sum_j a_{ij} E X_j + b_i = \sum_j a_{ij} \mu_j + b_i,$$

for each i. In vector notation, this is $E\mathbf{Y} = A\mu + \mathbf{b}$.

(ii) Write $\Sigma = (\sigma_{ij})$ for Σ_X. Then $Y_i - EY_i = \sum_k a_{ik}(X_k - EX_k) = \sum_k a_{ik}(X_k - \mu_k)$, so

$$
\begin{aligned}
\mathrm{cov}(Y_i, Y_j) &= E\left[\sum_r a_{ir}(X_r - \mu_r) \sum_s a_{js}(X_s - \mu_s)\right] \\
&= \sum_{rs} a_{ir} a_{js} E[(X_r - \mu_r)(X_s - \mu_s)] \\
&= \sum_{rs} a_{ir} a_{js} \sigma_{rs} = \sum_{rs} A_{ir} \Sigma_{rs} \left(A^T\right)_{sj} = \left(A\Sigma A^T\right)_{ij},
\end{aligned}
$$

identifying the elements of the matrix product $A\Sigma A^T$. $\qquad\square$

The same method of proof gives the following result, which we shall need later in connection with mixed models in §9.1.

Proposition 4.5

If $\mathbf{Z} = A\mathbf{X} + B\mathbf{Y}$ with constant matrices A, B and uncorrelated random vectors $\mathbf{X}$, $\mathbf{Y}$ with covariance matrices $\Sigma_{\mathbf{X}}$, $\Sigma_{\mathbf{Y}}$, $\mathbf{Z}$ has covariance matrix

$$\mathrm{cov}\,\mathbf{Z} = A\Sigma_{\mathbf{X}} A^T + B\Sigma_{\mathbf{Y}} B^T.$$

Corollary 4.6

Covariance matrices Σ are non-negative definite.

Proof

Let $\mathbf{a}$ be any $n \times 1$ matrix (row-vector of length n); then $Y := \mathbf{aX}$ is a scalar. So $Y = Y^T = \mathbf{Xa}^T$. Taking $\mathbf{a} = A^T, \mathbf{b} = \mathbf{0}$ above, Y has variance $\mathbf{a}^T \Sigma \mathbf{a}$ (a 1×1 covariance matrix). But variances are non-negative. So $\mathbf{a}^T \Sigma \mathbf{a} \geq \mathbf{0}$ for all n-vectors $\mathbf{a}$. This says that Σ is non-negative definite. □

We turn now to a technical result, which is important in reducing n-dimensional problems to one-dimensional ones.

Theorem 4.7 (Cramér-Wold device)

The distribution of a random n-vector $\mathbf{X}$ is completely determined by the set of all one-dimensional distributions of linear combinations $\mathbf{t}^T \mathbf{X} = \sum_i t_i X_i$, where $\mathbf{t}$ ranges over all fixed n-vectors.

Proof

When the MGF exists (as here), $Y := \mathbf{t}^T \mathbf{X}$ has MGF

$$M_Y(s) := E \exp\{sY\} = E \exp\{s\mathbf{t}^T \mathbf{X}\}.$$

If we know the distribution of each Y, we know its MGF $M_Y(s)$. In particular, taking $s = 1$, we know $E \exp\{\mathbf{t}^T \mathbf{X}\}$. But this is the MGF of $\mathbf{X} = (X_1, \ldots, X_n)^T$ evaluated at $\mathbf{t} = (t_1, \ldots, t_n)^T$. But this determines the distribution of $\mathbf{X}$.

When MGFs do not exist, replace $\mathbf{t}$ by $i\mathbf{t}$ ($i = \sqrt{-1}$) and use characteristic functions (CFs) instead. □

Thus by the Cramér–Wold device, to define an n-dimensional distribution it suffices to define the distributions of *all linear combinations*.

The Cramér–Wold device suggests a way to *define* the multivariate normal distribution. The definition below seems indirect, but it has the advantage of handling the full-rank and singular cases together ($\rho = \pm 1$ as well as $-1 < \rho < 1$ for the bivariate case).

Definition 4.8

An n-vector $\mathbf{X}$ has an *n-variate normal* distribution iff $\mathbf{a}^T \mathbf{X}$ has a univariate normal distribution for all constant n-vectors $\mathbf{a}$.

First, some properties resulting from the definition.

Proposition 4.9

(i) Any linear transformation of a multinormal n-vector is multinormal.

(ii) Any vector of elements from a multinormal n-vector is multinormal. In particular, the components are univariate normal.

Proof

(i) If $\mathbf{Y} = A\mathbf{X} + \mathbf{c}$ (A an $m \times n$ matrix, $\mathbf{c}$ an m-vector) is an m-vector, and $\mathbf{b}$ is any m-vector,

$$\mathbf{b}^T \mathbf{Y} = \mathbf{b}^T (A\mathbf{X} + \mathbf{c}) = (\mathbf{b}^T A)\mathbf{X} + \mathbf{b}^T \mathbf{c}.$$

If $\mathbf{a} = A^T \mathbf{b}$ (an m-vector), $\mathbf{a}^T \mathbf{X} = \mathbf{b}^T A\mathbf{X}$ is univariate normal as $\mathbf{X}$ is multinormal. Adding the constant $\mathbf{b}^T \mathbf{c}$, $\mathbf{b}^T \mathbf{Y}$ is univariate normal. This holds for all $\mathbf{b}$, so $\mathbf{Y}$ is m-variate normal.

(ii) Take a suitable matrix $\mathbf{A}$ of 1s and 0s to pick out the required sub-vector.
$\square$

Theorem 4.10

If $\mathbf{X}$ is n-variate normal with mean μ and covariance matrix Σ, its MGF is

$$M(\mathbf{t}) := E \exp\left\{\mathbf{t}^T \mathbf{X}\right\} = \exp\left\{\mathbf{t}^T \mu + \frac{1}{2}\mathbf{t}^T \Sigma \mathbf{t}\right\}.$$

Proof

By Proposition 4.4, $Y := \mathbf{t}^T \mathbf{X}$ has mean $\mathbf{t}^T \mu$ and variance $\mathbf{t}^T \Sigma \mathbf{t}$. By definition of multinormality, $Y = \mathbf{t}^T \mathbf{X}$ is univariate normal. So Y is $N(\mathbf{t}^T \mu, \mathbf{t}^T \Sigma \mathbf{t})$. So Y has MGF

$$M_Y(s) := E \exp\{sY\} = \exp\left\{s\mathbf{t}^T \mu + \frac{1}{2}s^2\mathbf{t}^T \Sigma \mathbf{t}\right\}.$$

But $E(e^{sY}) = E \exp\{s\mathbf{t}^T \mathbf{X}\}$, so taking $s = 1$ (as in the proof of the Cramér–Wold device),

$$E \exp\left\{\mathbf{t}^T \mathbf{X}\right\} = \exp\left\{\mathbf{t}^T \mu + \frac{1}{2}\mathbf{t}^T \Sigma \mathbf{t}\right\},$$

giving the MGF of $\mathbf{X}$ as required.
$\square$

Corollary 4.11

The components of $\mathbf{X}$ are independent iff Σ is diagonal.

Proof

The components are independent iff the joint MGF factors into the product of the marginal MGFs. This factorisation takes place, into $\prod_i \exp\{\mu_i t_i + \frac{1}{2}\sigma_{ii}t_i^2\}$, in the diagonal case only. $\qquad\qquad\square$

Recall that a covariance matrix Σ is always

(i) symmetric: ($\sigma_{ij} = \sigma_{ji}$, as $\sigma_{ij} = \text{cov}(X_i, X_j)$),

(ii) non-negative definite: $\mathbf{a}^T \Sigma \mathbf{a} \geq 0$ for all n-vectors $\mathbf{a}$.

Suppose that Σ is, further, *positive definite*:

$$\mathbf{a}^T \Sigma \mathbf{a} > 0 \qquad \text{unless} \qquad \mathbf{a} = \mathbf{0}.$$

(We write $\Sigma > 0$ for 'Σ is positive definite', $\Sigma \geq 0$ for 'Σ is non-negative definite'.)

Recall

a) a symmetric matrix has all its eigenvalues real (see e.g. Blyth and Robertson (2002b), Theorem 8.9, Corollary),

b) a symmetric non-negative definite matrix has all its eigenvalues non-negative (Blyth and Robertson (2002b) Th. 8.13),

c) a symmetric positive definite matrix is non-singular (has an inverse), and has all its eigenvalues positive (Blyth and Robertson (2002b), Th. 8.15).

We quote (see e.g. Halmos (1979), §79, Mardia, Kent and Bibby (1979)):

Theorem 4.12 (Spectral Decomposition)

If A is a symmetric matrix, A can be written

$$A = \Gamma \Lambda \Gamma^T,$$

where Λ is a diagonal matrix of eigenvalues of A and Γ is an orthogonal matrix whose columns are normalised eigenvectors.

Corollary 4.13

(i) For Σ a covariance matrix, we can define its *square root* matrix $\Sigma^{\frac{1}{2}}$ by
$$\Sigma^{\frac{1}{2}} := \Gamma \Lambda^{\frac{1}{2}} \Gamma^T, \ \Lambda^{\frac{1}{2}} := diag(\lambda_i^{\frac{1}{2}}), \text{ with } \Sigma^{\frac{1}{2}} \Sigma^{\frac{1}{2}} = \Sigma.$$

(ii) For Σ a non-singular (that is positive definite) covariance matrix, we can define its *inverse square root* matrix $\Sigma^{-\frac{1}{2}}$ by

$$\Sigma^{-\frac{1}{2}} := \Gamma\Lambda^{-\frac{1}{2}}\Gamma^T, \qquad \Lambda^{-\frac{1}{2}} := diag(\lambda^{-\frac{1}{2}}), \qquad \text{with} \qquad \Lambda^{-\frac{1}{2}}\Lambda^{-\frac{1}{2}} = \Lambda^{-1}.$$

Theorem 4.14

If X_i are independent (univariate) normal, any linear combination of the X_i is normal. That is, $\mathbf{X} = (X_1, \ldots, X_n)^T$, with X_i independent normal, is multinormal.

Proof

If X_i are independent $N(\mu_i, \sigma_i^2)$ $(i = 1, \ldots, n)$, $Y := \sum_i a_i X_i + c$ is a linear combination, Y has MGF

$$\begin{aligned}
M_Y(t) &:= E\exp\left\{t(c + \sum_i a_i X_i)\right\} \\
&= e^{tc}E\prod_i \exp\{ta_i X_i\} \qquad \text{(property of exponentials)} \\
&= e^{tc}E\prod_i E\exp\{ta_i X_i\} \qquad \text{(independence)} \\
&= e^{tc}\prod_i \exp\left\{\mu_i(a_i t) + \frac{1}{2}\sigma_i^2(a_i t)^2\right\} \qquad \text{(normal MGF)} \\
&= \exp\left\{\left[c + \sum_i a_i\mu_i\right]t + \frac{1}{2}\left[\sum_i a_i^2\sigma_i^2\right]t^2\right\},
\end{aligned}$$

so Y is $N(c + \sum_i a_i\mu_i, \sum_i a_i^2\sigma_i^2)$, from its MGF. $\qquad\qquad\square$

Independence of Linear Forms. Given a normally distributed random vector $\mathbf{x} \sim N(\mu, \Sigma)$ and a matrix A, one may form the *linear form* $A\mathbf{x}$. One often encounters several of these together, and needs their joint distribution – in particular, to know when these are independent.

Theorem 4.15

Linear forms $A\mathbf{x}$ and $B\mathbf{x}$ with $\mathbf{x} \sim N(\mu, \Sigma)$ are independent iff

$$A\Sigma B^T = 0.$$

In particular, if A, B are symmetric and $\Sigma = \sigma^2 I$, they are independent iff

$$AB = 0.$$

Proof

The joint MGF is

$$M(\mathbf{u}, \mathbf{v}) := E \exp\left\{\mathbf{u}^T A \mathbf{x} + \mathbf{v}^T B \mathbf{x}\right\} = E \exp\left\{\left(A^T \mathbf{u} + B^T \mathbf{v}\right)^T \mathbf{x}\right\}.$$

This is the MGF of $\mathbf{x}$ at argument $\mathbf{t} = A^T \mathbf{u} + B^T \mathbf{v}$, so $M(\mathbf{u}, \mathbf{v})$ is given by

$$\exp\{(\mathbf{u}^T A + \mathbf{v}^T B)\mu + \frac{1}{2}[\mathbf{u}^T A \Sigma A^T \mathbf{u} + \mathbf{u}^T A \Sigma B^T \mathbf{v} + \mathbf{v}^T B \Sigma A^T \mathbf{u} + \mathbf{v}^T B \Sigma B^T \mathbf{v}]\}.$$

This factorises into a product of a function of $\mathbf{u}$ and a function of $\mathbf{v}$ iff the two cross terms in $\mathbf{u}$ and $\mathbf{v}$ vanish, that is, iff $A \Sigma B^T = 0$ and $B \Sigma A^T = 0$; by symmetry of Σ, the two are equivalent. $\qquad\square$

4.4 The Multinormal Density

If $\mathbf{X}$ is n-variate normal, $N(\mu, \Sigma)$, its density (in n dimensions) need not exist (e.g. the singular case $\rho = \pm 1$ with $n = 2$ in Chapter 1). But if $\Sigma > 0$ (so Σ^{-1} exists), $\mathbf{X}$ has a density. The link between the multinormal density below and the multinormal MGF above is due to the English statistician F. Y. Edgeworth (1845–1926).

Theorem 4.16 (Edgeworth's Theorem, 1893)

If μ is an n-vector, $\Sigma > 0$ a symmetric positive definite $n \times n$ matrix, then

(i)

$$f(\mathbf{x}) := \frac{1}{(2\pi)^{\frac{1}{2}n}|\Sigma|^{\frac{1}{2}}} \exp\left\{-\frac{1}{2}(\mathbf{x} - \mu)^T \Sigma^{-1}(\mathbf{x} - \mu)\right\}$$

is an n-dimensional probability density function (of a random n-vector $\mathbf{X}$, say),

(ii) $\mathbf{X}$ has MGF $M(\mathbf{t}) = \exp\left\{\mathbf{t}^T \mu + \frac{1}{2}\mathbf{t}^T \Sigma \mathbf{t}\right\}$,

(iii) $\mathbf{X}$ is multinormal $N(\mu, \Sigma)$.

Proof

Write $\mathbf{Y} := \Sigma^{-\frac{1}{2}}\mathbf{X}$ ($\Sigma^{-\frac{1}{2}}$ exists as $\Sigma > 0$, by above). Then $\mathbf{Y}$ has covariance matrix $\Sigma^{-\frac{1}{2}}\Sigma(\Sigma^{-\frac{1}{2}})^T$. Since $\Sigma = \Sigma^T$ and $\Sigma = \Sigma^{\frac{1}{2}}\Sigma^{\frac{1}{2}}$, $\mathbf{Y}$ has covariance matrix I (the components Y_i of $\mathbf{Y}$ are uncorrelated).

Change variables as above, with $\mathbf{y} = \Sigma^{-\frac{1}{2}}\mathbf{x}$, $\mathbf{x} = \Sigma^{\frac{1}{2}}\mathbf{y}$. The Jacobian is (taking $A = \Sigma^{-\frac{1}{2}}$) $J = \partial\mathbf{x}/\partial\mathbf{y} = det(\Sigma^{\frac{1}{2}}), = (det\Sigma)^{\frac{1}{2}}$ by the product theorem for determinants. Substituting,

$$\exp\left\{-\frac{1}{2}(\mathbf{x}-\mu)^T\Sigma^{-1}(\mathbf{x}-\mu)\right\}$$

is

$$\exp\left\{-\frac{1}{2}\left(\Sigma^{\frac{1}{2}}\mathbf{y} - \Sigma^{\frac{1}{2}}\left(\Sigma^{-\frac{1}{2}}\mu\right)\right)^T\Sigma^{-1}\left(\Sigma^{\frac{1}{2}}\mathbf{y} - \Sigma^{\frac{1}{2}}\left(\Sigma^{-\frac{1}{2}}\mu\right)\right)\right\},$$

or writing $\nu := \Sigma^{-\frac{1}{2}}\mu$,

$$\exp\left\{-\frac{1}{2}(\mathbf{y}-\nu)^T\Sigma^{\frac{1}{2}}\Sigma^{-1}\Sigma^{\frac{1}{2}}(\mathbf{y}-\nu)\right\} = \exp\left\{-\frac{1}{2}(\mathbf{y}-\nu)^T(\mathbf{y}-\nu)\right\}.$$

So by the change of density formula, $\mathbf{Y}$ has density $g(\mathbf{y})$ given by

$$\frac{1}{(2\pi)^{\frac{1}{2}n}|\Sigma|^{\frac{1}{2}}}|\Sigma|^{\frac{1}{2}}\exp\{-\frac{1}{2}(\mathbf{y}-\nu)^T(\mathbf{y}-\nu)\} = \prod_{i=1}^{n}\frac{1}{(2\pi)^{\frac{1}{2}}}\exp\{-\frac{1}{2}(y_i-\nu_i)^2\}.$$

This is the density of a multivariate vector $y \sim N(\nu, I)$ whose components are independent $N(\nu_i, 1)$ by Theorem 4.14.

(i) Taking $A = B = \mathbb{R}^n$ in the Jacobian formula,

$$\begin{aligned}
\int_{\mathbb{R}^n} f(\mathbf{x})d\mathbf{x} &= \frac{1}{(2\pi)^{\frac{1}{2}n}}|\Sigma|^{\frac{1}{2}}\int_{\mathbb{R}^n}\exp\left\{-\frac{1}{2}(\mathbf{x}-\mu)^T\Sigma^{-1}(\mathbf{x}-\mu)\right\}d\mathbf{x} \\
&= \frac{1}{(2\pi)^{\frac{1}{2}n}}\int_{\mathbb{R}^n}\exp\left\{-\frac{1}{2}(\mathbf{y}-\nu)^T(\mathbf{y}-\nu)\right\}d\mathbf{y} \\
&= \int_{\mathbb{R}^n} g(\mathbf{y})d\mathbf{y} = 1.
\end{aligned}$$

So $f(\mathbf{x})$ is a probability density (of $\mathbf{X}$ say).

(ii) $\mathbf{X} = \Sigma^{\frac{1}{2}}\mathbf{Y}$ is a linear transformation of $\mathbf{Y}$, and $\mathbf{Y}$ is multivariate normal, so $\mathbf{X}$ is multivariate normal.

(iii) $E\mathbf{X} = \Sigma^{\frac{1}{2}}E\mathbf{Y} = \Sigma^{\frac{1}{2}}\nu = \Sigma^{\frac{1}{2}}.\Sigma^{-\frac{1}{2}}\mu = \mu$, $\text{cov}\mathbf{X} = \Sigma^{\frac{1}{2}}\text{cov}\mathbf{Y}(\Sigma^{\frac{1}{2}})^T = \Sigma^{\frac{1}{2}}I\Sigma^{\frac{1}{2}} = \Sigma$. So $\mathbf{X}$ is multinormal $N(\mu, \Sigma)$. So its MGF is

$$M(\mathbf{t}) = \exp\left\{\mathbf{t}^T\mu + \frac{1}{2}\mathbf{t}^T\Sigma\mathbf{t}\right\}. \qquad \square$$

4.4.1 Estimation for the multivariate normal

Given a sample $x_1, \ldots, x_n$ from the multivariate normal $N_p(\mu, \Sigma)$, $\Sigma > 0$, form the *sample mean* (vector)

$$\overline{x} := \frac{1}{n} \sum_{i=1}^n x_i,$$

as in the one-dimensional case, and the *sample covariance matrix*

$$S := \frac{1}{n} \sum_{i=1}^n (x_i - \overline{x})(x_i - \overline{x})^T.$$

The likelihood for a sample of size 1 is

$$L = (2\pi)^{-p/2} |\Sigma|^{-1/2} \exp\left\{ -\frac{1}{2}(x - \mu)^T \Sigma^{-1}(x - \mu) \right\},$$

so the likelihood for a sample of size n is

$$L = (2\pi)^{-np/2} |\Sigma|^{-n/2} \exp\left\{ -\frac{1}{2} \sum_{1}^n (x_i - \mu)^T \Sigma^{-1}(x_i - \mu) \right\}.$$

Writing

$$x_i - \mu = (x_i - \overline{x}) - (\mu - \overline{x}),$$

$$\sum_{1}^n (x_i - \mu)^T \Sigma^{-1}(x_i - \mu) = \sum_{1}^n (x_i - \overline{x})^T \Sigma^{-1}(x_i - \overline{x}) + n(\overline{x} - \mu)^T \Sigma^{-1}(\overline{x} - \mu)$$

(the cross terms cancel as $\sum(x_i - \overline{x}) = 0$). The summand in the first term on the right is a scalar, so is its own trace. Since $\text{trace}(AB) = \text{trace}(BA)$ and $\text{trace}(A + B) = \text{trace}(A) + \text{trace}(B) = \text{trace}(B + A)$,

$$\text{trace}\left(\sum_{1}^n (x_i - \overline{x})^T \Sigma^{-1}(x_i - \overline{x}) \right) = \text{trace}\left(\Sigma^{-1} \sum_{1}^n (x_i - \overline{x})(x_i - \overline{x})^T \right)$$

$$= \text{trace}\left(\Sigma^{-1}.nS \right) = n \text{ trace}\left(\Sigma^{-1}S \right).$$

Combining,

$$L = (2\pi)^{-np/2} |\Sigma|^{-n/2} \exp\left\{ -\frac{1}{2}n \text{ trace}\left(\Sigma^{-1}S \right) - \frac{1}{2}n(\overline{x} - \mu)^T \Sigma^{-1}(\overline{x} - \mu) \right\}.$$

This involves the data only through $\overline{x}$ and S. We expect the sample mean $\overline{x}$ to be informative about the population mean μ and the sample covariance matrix S to be informative about the population covariance matrix Σ. In fact $\overline{x}$, S are fully informative about μ, Σ, in a sense that can be made precise using the theory of *sufficient statistics* (for which we must refer to a good book on statistical inference – see e.g. Casella and Berger (1990), Ch. 6) – another of Fisher's contributions. These natural estimators are in fact the maximum likelihood estimators:

Theorem 4.17

For the multivariate normal $N_p(\mu, \Sigma)$, $\bar{x}$ and S are the maximum likelihood estimators for μ, Σ.

Proof

Write $V = (v_{ij}) := \Sigma^{-1}$. By above, the likelihood is

$$L = \text{const.}|V|^{n/2} \exp\left\{ -\frac{1}{2}n \, \text{trace}(VS) - \frac{1}{2}n(\bar{x} - \mu)^T V(\bar{x} - \mu) \right\},$$

so the log-likelihood is

$$\ell = c + \frac{1}{2}n \log|V| - \frac{1}{2}n \, \text{trace}(VS) - \frac{1}{2}n(\bar{x} - \mu)^T V(\bar{x} - \mu).$$

The MLE $\hat{\mu}$ for μ is $\bar{x}$, as this reduces the last term (the only one involving μ) to its minimum value, 0. Recall (see e.g. Blyth and Robertson (2002a), Ch. 8) that for a square matrix $A = (a_{ij})$, its determinant is

$$|A| = \sum_j a_{ij} A_{ij}$$

for each i, or

$$|A| = \sum_i a_{ij} A_{ij}$$

for each j, expanding by the ith row or jth column, where A_{ij} is the *cofactor* (signed minor) of a_{ij}. From either,

$$\partial|A|/\partial a_{ij} = A_{ij},$$

so

$$\partial \log|A|/\partial a_{ij} = A_{ij}/|A| = (A^{-1})_{ij},$$

the (i, j) element of A^{-1}, recalling the formula for the matrix inverse. Also, if B is symmetric,

$$\text{trace}(AB) = \sum_i \sum_j a_{ij} b_{ji} = \sum_{i,j} a_{ij} b_{ij},$$

so

$$\partial \, \text{trace}(AB)/\partial a_{ij} = b_{ij}.$$

Using these, and writing $S = (s_{ij})$,

$$\partial \log|V|/\partial v_{ij} = (V^{-1})_{ij} = (\Sigma)_{ij} = \sigma_{ij} \qquad (V := \Sigma^{-1}),$$

$$\partial \, \text{trace}(VS)/\partial v_{ij} = s_{ij}.$$

So

$$\partial \ell/\partial v_{ij} = \frac{1}{2}n(\sigma_{ij} - s_{ij}),$$

which is 0 for all i and j iff $\Sigma = S$. This says that S is the MLE for Σ, as required. $\qquad\square$

4.5 Conditioning and Regression

Recall from §1.5 that the *conditional* density of Y *given* $X = x$ is

$$f_{Y|X}(y|x) := f_{X,Y}(x,y)/\int f_{X,Y}(x,y)\ dy.$$

Conditional means. The conditional mean of Y given $X = x$ is

$$E(Y|X = x),$$

a function of x called the *regression* function (of Y on x). So, if we do not specify the value x, we get $E(Y|X)$. This is *random*, because X is random (until we observe its value, x; then we get the regression function of x as above). As $E(Y|X)$ is random, we can look at its mean and variance. For the next result, see e.g. Haigh (2002) Th. 4.24 or Williams (2001), §9.1.

Theorem 4.18 (Conditional Mean Formula)

$$E[E(Y|X)] = EY.$$

Proof

$$
\begin{aligned}
EY &= \int y f_Y(y)dy = \int y dy \int f_{X,Y}(x,y)\ dx \\
&= \int y\ dy \int f_{Y|X}(y|x)f_X(x)\ dx \qquad \text{(definition of conditional density)} \\
&= \int f_X(x)\ dx \int y f_{Y|X}(y|x)\ dx,
\end{aligned}
$$

interchanging the order of integration. The inner integral is $E(Y|X = x)$. The outer integral takes the expectation of this over X, giving $E[E(Y|X)]$.

Discrete case: similarly with summation in place of integration.   □

Interpretation.

- EY takes the random variable Y, and averages out all the randomness to give a number, EY.

- $E(Y|X)$ takes the random variable Y knowing X, and averages out all the randomness in Y NOT accounted for by knowledge of X.

– $E[E(Y|X)]$ then averages out the remaining randomness, which IS accounted for by knowledge of X, to give EY as above.

Example 4.19 (Bivariate normal distribution)

$N(\mu_1, \mu_2; \sigma_1^2, \sigma_2^2; \rho)$, or $N(\mu, \Sigma)$,

$$\mu = (\mu_1, \mu_2)^T, \qquad \Sigma = \begin{pmatrix} \sigma_1^2 & \rho\sigma_1\sigma_2 \\ \rho\sigma_1\sigma_2 & \sigma_2^2 \end{pmatrix} = \begin{pmatrix} \sigma_{11}^2 & \sigma_{12} \\ \sigma_{12} & \sigma_{22}^2 \end{pmatrix}.$$

By §1.5,

$$E(Y|X = x) = \mu_2 + \rho\frac{\sigma_2}{\sigma_1}(x - \mu_1), \qquad \text{so} \qquad E(Y|X) = \mu_2 + \rho\frac{\sigma_2}{\sigma_1}(X - \mu_1).$$

So

$$E[E(Y|X)] = \mu_2 + \rho\frac{\sigma_2}{\sigma_1}(EX - \mu_1) = \mu_2 = EY, \qquad \text{as} \qquad EX = \mu_1.$$

As with the bivariate normal, we should keep some concrete instance in mind as a motivating example, e.g.:

X = incoming score of student [in medical school or university, say], Y = graduating score;

X = child's height at 2 years (say), Y = child's eventual adult height,

or

X = mid-parental height, Y = child's adult height, as in Galton's study.

Conditional variances. Recall $\mathrm{var}X := E[(X - EX)^2]$. Expanding the square,

$$\begin{aligned} \mathrm{var}X &= E\left[X^2 - 2X.(EX) + (EX)^2\right] = E\left(X^2\right) - 2(EX)(EX) + (EX)^2, \\ &= E\left(X^2\right) - (EX)^2. \end{aligned}$$

Conditional variances can be defined in the same way. Recall that $E(Y|X)$ is constant when X is known ($= x$, say), so can be taken outside an expectation over X, E_X say. Then

$$\mathrm{var}(Y|X) := E(Y^2|X) - [E(Y|X)]^2.$$

Take expectations of both sides over X:

$$E_X[\mathrm{var}(Y|X)] = E_X[E(Y^2|X)] - E_X[E(Y|X)]^2.$$

Now $E_X[E(Y^2|X)] = E(Y^2)$, by the Conditional Mean Formula, so the right is, adding and subtracting $(EY)^2$,

$$\{E(Y^2) - (EY)^2\} - \{E_X[E(Y|X)]^2 - (EY)^2\}.$$

The first term is var Y, by above. Since $E(Y|X)$ has E_X-mean EY, the second term is $\mathrm{var}_X E(Y|X)$, the variance (over X) of the random variable $E(Y|X)$ (random because X is). Combining, we have (Williams (2001), §9.1, or Haigh (2002) Ex 4.33):

Theorem 4.20 (Conditional Variance Formula)

$$\mathrm{var}Y = E_X \mathrm{var}(Y|X) + \mathrm{var}_X E(Y|X).$$

Interpretation.

– $\mathrm{var}Y$ = total variability in Y,

– $E_X \mathrm{var}(Y|X)$ = variability in Y not accounted for by knowledge of X,

– $\mathrm{var}_X E(Y|X)$ = variability in Y accounted for by knowledge of X.

Example 4.21 (The Bivariate normal)

$$Y|X = x \text{ is } N\left(\mu_2 + \rho\frac{\sigma_2}{\sigma_1}(x - \mu_1), \sigma_2^2\left(1 - \rho^2\right)\right), \qquad \mathrm{var}Y = \sigma_2^2,$$

$$E(Y|X = x) = \mu_2 + \rho\frac{\sigma_2}{\sigma_1}(x - \mu_1), \qquad E(Y|X) = \mu_2 + \rho\frac{\sigma_2}{\sigma_1}(X - \mu_1),$$

which has variance

$$\mathrm{var}\, E(Y|X) = (\rho\sigma_2/\sigma_1)^2 \mathrm{var}X = (\rho\sigma_2/\sigma_1)^2\sigma_1^2 = \rho^2\sigma_2^2,$$

$$\mathrm{var}(Y|X = x) = \sigma_2^2\left(1 - \rho^2\right) \text{ for all } x, \quad \mathrm{var}(Y|X) = \sigma_2^2\left(1 - \rho^2\right),$$

(as in Fact 6 of §1.5):

$$E_X \mathrm{var}(Y|X) = \sigma_2^2\left(1 - \rho^2\right).$$

Corollary 4.22

$E(Y|X)$ has the same mean as Y and smaller variance (if anything) than Y.

Proof

From the Conditional Mean Formula, $E[E(Y|X)] = EY$. Since $\mathrm{var}(Y|X) \geq 0$, $E_X \mathrm{var}(Y|X) \geq 0$, so

$$\mathrm{var}E[Y|X] \leq \mathrm{var}Y$$

from the Conditional Variance Formula.   □

Note 4.23

This result has important applications in estimation theory. Suppose we are to estimate a parameter θ, and are considering a statistic X as a possible estimator (or basis for an estimator) of θ. We would naturally want X to contain all the information on θ contained within the entire sample. What (if anything) does this mean in precise terms? The answer lies in Fisher's concept of *sufficiency* ('data reduction'), that we met in §4.4.1. In the language of sufficiency, the Conditional Variance Formula is seen as (essentially) the Rao–Blackwell Theorem, a key result in the area.

Regression. In the bivariate normal, with $X =$ mid-parent height, $Y =$ child's height, $E(Y|X = x)$ is linear in x (*regression line*). In a more detailed analysis, with $U =$ father's height, $V =$ mother's height, $Y =$ child's height, one would expect $E(Y|U = u, V = v)$ to be linear in u and v (*regression plane*), etc.

In an n-variate normal distribution $N_n(\mu, \Sigma)$, suppose that we partition $\mathbf{X} = (X_1, \ldots, X_n)^T$ into $\mathbf{X}_1 := (X_1, \ldots, X_r)^T$ and $\mathbf{X}_2 := (X_{r+1}, \ldots, X_n)^T$. Let the corresponding partition of the mean vector and the covariance matrix be

$$\mu = \left(\begin{array}{c} \mu_1 \\ \mu_2 \end{array} \right), \quad \Sigma = \left(\begin{array}{cc} \Sigma_{11} & \Sigma_{12} \\ \Sigma_{21} & \Sigma_{22} \end{array} \right),$$

where $E\mathbf{X}_i = \mu_i$, Σ_{11} is the covariance matrix of $\mathbf{X}_1$, Σ_{22} that of $\mathbf{X}_2$, $\Sigma_{12} = \Sigma_{21}^T$ the covariance matrix of $\mathbf{X}_1$ with $\mathbf{X}_2$.

For clarity, we restrict attention to the non-singular case, where Σ is positive definite.

Lemma 4.24

If Σ is positive definite, so is Σ_{11}.

Proof

$\mathbf{x}^T \Sigma \mathbf{x} > \mathbf{0}$ as Σ is positive definite. Take $\mathbf{x} = (\mathbf{x}_1, \mathbf{0})^T$, where $\mathbf{x}_1$ has the same number of components as the order of Σ_{11} (that is, in matrix language, so that the partition of $\mathbf{x}$ is conformable with those of μ and Σ above). Then $\mathbf{x}_1 \Sigma_{11} \mathbf{x}_1 > 0$ for all $\mathbf{x}_1$. This says that Σ_{11} is positive definite, as required. $\square$

Theorem 4.25

The conditional distribution of $\mathbf{X}_2$ given $\mathbf{X}_1 = \mathbf{x}_1$ is

$$\mathbf{X}_2|\mathbf{X}_1 = \mathbf{x}_1 \sim N\left(\mu_2 + \Sigma_{21}\Sigma_{11}^{-1}(\mathbf{x}_1 - \mu_1), \Sigma_{22} - \Sigma_{21}\Sigma_{11}^{-1}\Sigma_{12}\right).$$

Corollary 4.26

The regression of $\mathbf{X}_2$ on $\mathbf{X}_1$ is linear:

$$E(\mathbf{X}_2|\mathbf{X}_1 = \mathbf{x}_1) = \mu_2 + \Sigma_{21}\Sigma_{11}^{-1}(\mathbf{x}_1 - \mu_1).$$

Proof

Recall from Theorem 4.16 that $A\mathbf{X}, B\mathbf{X}$ are independent iff $A\Sigma B^T = \mathbf{0}$, or as Σ is symmetric, $B\Sigma A^T = \mathbf{0}$. Now

$$\mathbf{X}_1 = A\mathbf{X} \text{ where } A = (I, 0),$$

$$\mathbf{X}_2 - \Sigma_{21}\Sigma_{11}^{-1}\mathbf{X}_1 = \left(-\Sigma_{21}\Sigma_{11}^{-1} \ \ I\right)\begin{pmatrix}\mathbf{X}_1 \\ \mathbf{X}_2\end{pmatrix} = B\mathbf{X}, \text{ where } B = \left(-\Sigma_{21}\Sigma_{11}^{-1} \ \ I\right).$$

Now

$$
\begin{aligned}
B\Sigma A^T &= \left(-\Sigma_{21}\Sigma_{11}^{-1} \ \ I\right)\begin{bmatrix} \Sigma_{11} & \Sigma_{12} \\ \Sigma_{21} & \Sigma_{22} \end{bmatrix}\begin{bmatrix} I \\ 0 \end{bmatrix} \\
&= \left(-\Sigma_{21}\Sigma_{11}^{-1} \ \ I\right)\begin{bmatrix} \Sigma_{11} \\ \Sigma_{21} \end{bmatrix} \\
&= -\Sigma_{21}\Sigma_{11}^{-1}\Sigma_{11} + \Sigma_{21} = 0,
\end{aligned}
$$

so $\mathbf{X}_1$ and $\mathbf{X}_2 - \Sigma_{21}\Sigma_{11}^{-1}\mathbf{X}_1$ are *independent*. Since both are linear transformations of $\mathbf{X}$, which is multinormal, both are *multinormal*. Also,

$$E(B\mathbf{X}) = BE\mathbf{X} = \left(-\Sigma_{21}\Sigma_{11}^{-1} \ \ I\right)\begin{pmatrix}\mu_1 & \mu_2\end{pmatrix} = \mu_2 - \Sigma_{21}\Sigma_{11}^{-1}\mu_1.$$

To calculate the covariance matrix, introduce $C := -\Sigma_{21}\Sigma_{11}^{-1}$, so $B = (C \ I)$, and recall $\Sigma_{12}^T = \Sigma_{21}$, so $C^T = -\Sigma_{11}^{-1}\Sigma_{12}$:

$$\text{var}(B\mathbf{X}) = B\Sigma B^T = \left(C \ \ I\right)\begin{bmatrix} \Sigma_{11} & \Sigma_{12} \\ \Sigma_{21} & \Sigma_{22} \end{bmatrix}\begin{bmatrix} C^T \\ I \end{bmatrix}$$

$$= \left(C \ \ I\right)\begin{bmatrix} \Sigma_{11}C^T + \Sigma_{12} \\ \Sigma_{21}C^T + \Sigma_{22} \end{bmatrix} = C\Sigma_{11}C^T + C\Sigma_{12} + \Sigma_{21}C^T + \Sigma_{22}$$

$$= \Sigma_{21}\Sigma_{11}^{-1}\Sigma_{11}\Sigma_{11}^{-1}\Sigma_{12} - \Sigma_{21}\Sigma_{11}^{-1}\Sigma_{12} - \Sigma_{21}\Sigma_{11}^{-1}\Sigma_{12} + \Sigma_{22}$$

$$= \Sigma_{22} - \Sigma_{21}\Sigma_{11}^{-1}\Sigma_{12}.$$

By independence, the conditional distribution of $B\mathbf{X}$ given $\mathbf{X}_1 = A\mathbf{X}$ is the same as its marginal distribution, which by above is $N(\mu_2 - \Sigma_{21}\Sigma_{11}^{-1}\mu_1, \Sigma_{22} - \Sigma_{21}\Sigma_{11}^{-1}\Sigma_{12})$. So given $\mathbf{X}_1$, $\mathbf{X}_2 - \Sigma_{21}\Sigma_{11}^{-1}\mathbf{X}_1$ is $N(\mu_2 - \Sigma_{21}\Sigma_{11}^{-1}\mu_1, \Sigma_{22} - \Sigma_{21}\Sigma_{11}^{-1}\Sigma_{12})$.

It remains to pass from the conditional distribution of $\mathbf{X}_2 - \Sigma_{21}\Sigma_{11}^{-1}\mathbf{X}_1$ given $\mathbf{X}_1$ to that of $\mathbf{X}_2$ given $\mathbf{X}_1$. But given $\mathbf{X}_1$, $\Sigma_{21}\Sigma_{11}^{-1}\mathbf{X}_1$ is constant, so we can do this simply by adding $\Sigma_{21}\Sigma_{11}^{-1}\mathbf{X}_1$. The result is again multinormal, with the same covariance matrix, but (conditional) mean $\mu_2 + \Sigma_{21}\Sigma_{11}^{-1}(\mathbf{X}_1 - \mu_1)$. That is, the conditional distribution of $\mathbf{X}_2$ given $\mathbf{X}_1$ is

$$N\left(\mu_2 + \Sigma_{21}\Sigma_{11}^{-1}(\mathbf{X}_1 - \mu_1), \Sigma_{22} - \Sigma_{21}\Sigma_{11}^{-1}\Sigma_{12}\right),$$

as required. $\square$

Note 4.27

Here $\Sigma_{22} - \Sigma_{21}\Sigma_{11}^{-1}\Sigma_{12}$ is called the *partial covariance matrix* of $\mathbf{X}_2$ given $\mathbf{X}_1$. In the language of Linear Algebra, it is called the Schur complement of Σ_{22} in Σ (Issai Schur (1875–1941) in 1905; see Zhang (2005)). We will meet the Schur complement again in §9.1 (see also Exercise 4.10).

Example 4.28 (Bivariate normal)

Here $n = 2, r = s = 1$:

$$\Sigma = \begin{pmatrix} \sigma_1^2 & \rho\sigma_1\sigma_2 \\ \rho\sigma_1\sigma_2 & \sigma_2^2 \end{pmatrix} = \begin{pmatrix} \Sigma_{11} & \Sigma_{12} \\ \Sigma_{21} & \Sigma_{22} \end{pmatrix},$$

$$\Sigma_{21}\Sigma_{11}^{-1}(X_1 - \mu_1) = \frac{\rho\sigma_1\sigma_2}{\sigma_1^2}(X_1 - \mu_1) = \frac{\rho\sigma_2}{\sigma_1}(X_1 - \mu_1),$$

$$\Sigma_{22} - \Sigma_{21}\Sigma_{11}^{-1}\Sigma_{12} = \sigma_2^2 - \rho\sigma_1\sigma_2.\sigma_1^{-2}.\rho\sigma_1\sigma_2 = \sigma_2^2(1 - \rho^2),$$

as before.

Note 4.29

The argument can be extended to cover the singular case as well as the non-singular case, using *generalised inverses* of the relevant matrices. For details, see e.g. Rao (1973), §8a.2v, 522–523.

Note 4.30

The details of the matrix algebra are less important than the result: *conditional distributions of multinormals are multinormal.* To find out *which* multinormal, we then only need to get the first and second moments – mean vector and covariance matrix – right.

Note 4.31

The result can actually be generalised well beyond the multivariate normal case. Recall (bivariate normal, Fact 8) that the bivariate normal has *elliptical contours*. The same is true in the multivariate normal case, by Edgeworth's Theorem – the contours are $Q(\mathbf{x}) := (\mathbf{x} - \mu)^T \Sigma^{-1}(\mathbf{x} - \mu) = $ constant. It turns out that this is the crucial property. *Elliptically contoured* distributions are much more general than the multivariate normal but share most of its nice properties, including having linear regression.

4.6 Mean-square prediction

Chapters 3 and 4 deal with *linear* prediction, but some aspects are more general. Suppose that y is to be predicted from a vector $\mathbf{x}$, by some predictor $f(\mathbf{x})$. One obvious candidate is the *regression function*

$$M(\mathbf{x}) := E[y|\mathbf{x}],$$

('M for mean'). Then

$$E[(y - M(\mathbf{x}))(M(\mathbf{x}) - f(\mathbf{x}))] = E[E[(y - M(\mathbf{x}))(M(\mathbf{x}) - f(\mathbf{x}))|\mathbf{x}]],$$

by the Conditional Mean Formula. But given $\mathbf{x}$, $M(\mathbf{x}) - f(\mathbf{x})$ is known, so can be taken through the inner expectation sign (like a constant). So the right is

$$E[(M(\mathbf{x}) - f(\mathbf{x}))E[(y - M(\mathbf{x}))|\mathbf{x}]].$$

But the inner expression is 0, as $M = E(y|\mathbf{x})$. So

$$
\begin{aligned}
E\left[(y - f)^2\right] &= E\left[((y - M) + (M - f))^2\right] \\
&= E\left[(y - M)^2\right] + 2E[(y - M)(M - f)] + E\left[(M - f)^2\right] \\
&= E\left[(y - M)^2\right] + E\left[(M - f)^2\right],
\end{aligned}
$$

by above. Interpreting the left as the mean-squared error – in brief, prediction error – when predicting y by $f(\mathbf{x})$, this says:

(i) $E[(y - M)^2] \leq E[(y - f)^2]$: M has prediction error at most that of f.

(ii) The regression function $M(\mathbf{x}) = E[y|\mathbf{x}]$ minimises the prediction error over all predictors f.

Now

$$
\begin{aligned}
\mathrm{cov}(y, f) &= E[(f - EF)(y - Ey)] \quad \text{(definition of covariance)} \\
&= E[(f - Ef)E[(y - Ey)|\mathbf{x}]] \quad \text{(Conditional Mean Formula)} \\
&= E[(f - Ef)(M - EM)] \quad \text{(definition of } M) \\
&= \mathrm{cov}(M, f).
\end{aligned}
$$

So

$$
\mathrm{corr}^2(f, y) = \frac{\mathrm{cov}^2(f, y)}{\mathrm{var}\, f\, \mathrm{var}\, y} = \frac{\mathrm{cov}^2(f, y)}{\mathrm{var} f \mathrm{var} M} \cdot \frac{\mathrm{var} M}{\mathrm{var}\, y} = \mathrm{corr}^2(M, f). \frac{\mathrm{var} M}{\mathrm{var}\, y}.
$$

When the predictor f is M, one has by above

$$
\mathrm{cov}(y, M) = \mathrm{cov}(M, M) = \mathrm{var} M.
$$

So

$$
\mathrm{corr}^2(y, M) = \frac{\mathrm{cov}^2(y, M)}{\mathrm{var}\, y\, \mathrm{var}\, M} = \frac{\mathrm{var}\, M}{\mathrm{var}\, y}.
$$

Combining,

$$
\mathrm{corr}^2(f, y) = \mathrm{corr}^2(f, M).\mathrm{corr}^2(M, y).
$$

Since correlation coefficients lie in $[-1, 1]$, and so their squares lie in $[0, 1]$, this gives

$$
\mathrm{corr}^2(f, y) \le \mathrm{corr}^2(M, y),
$$

with equality iff

$$
f = M.
$$

This gives

Theorem 4.32

The regression function $M(\mathbf{x}) := E(y|\mathbf{x})$ has the maximum squared correlation with y over all predictors $f(\mathbf{x})$ of y.

Note 4.33

1. One often uses the alternative notation $\rho(\cdot, \cdot)$ for the correlation $\mathrm{corr}(\cdot, \cdot)$. One then interprets $\rho^2 = \rho^2(M, y)$ as a measure of how well the regression M explains the data y.

2. The simplest example of this is the bivariate normal distribution of §1.5.

3. This interpretation of ρ^2 reinforces that it is the population counterpart of R^2 and its analogous interpretation in Chapter 3.

4. Since $\text{corr}^2(y, M) \leq 1$, one sees again that var $M \leq$ var y, as in the Conditional Variance Formula and the Rao–Blackwell Theorem, Theorem 4.20, Corollary 4.22 and Note 4.23.

5. This interpretation of regression as maximal correlation is another way of looking at regression in terms of projection, as in §3.6. For another treatment see Williams (2001), Ch. 8.

4.7 Generalised least squares and weighted regression

Suppose that we write down the model equation

$$\mathbf{y} = X\beta + \epsilon, \tag{GLS}$$

where it is assumed that

$$\epsilon \sim N(0, \sigma^2 V),$$

with $V \neq I$ in general. We take V full rank; then V^{-1} exists, $X^T V^{-1} X$ is full rank, and $(X^T V^{-1} X)^{-1}$ exists. (GLS) is the model equation for *generalised least squares*. If V is diagonal (GLS) is known as *weighted least squares*. By Corollary 4.13 (Matrix square roots) we can find P non-singular and symmetric such that

$$P^T P = P^2 = V.$$

Theorem 4.34 (Generalised Least Squares)

Under generalised least squares (GLS), the maximum likelihood estimate $\hat{\beta}$ of β is

$$\hat{\beta} = \left(X^T V^{-1} X \right)^{-1} X^T V^{-1} y.$$

This is also the best linear unbiased estimator (BLUE).

Proof

Pre–multiply by P^{-1} to reduce the equation for generalised least squares to the equation for ordinary least squares:

$$P^{-1}\mathbf{y} = P^{-1}X\beta + P^{-1}\epsilon. \qquad (OLS)$$

Now by Proposition 4.4 (ii)

$$\mathrm{cov}(P^{-1}\epsilon) = P^{-1}\mathrm{cov}(\epsilon)(P^{-1})^T = P^{-1}\sigma^2 V P^{-1} = \sigma^2.P^{-1}PPP^{-1} = \sigma^2 I.$$

So (OLS) is now a regression problem for β within the framework of ordinary least squares. From Theorem 3.5 the maximum likelihood estimate of β can now be obtained from the normal equations as

$$\begin{aligned}
\left[\left(P^{-1}X\right)^T \left(P^{-1}X\right)\right]^{-1}\left(P^{-1}X\right)^T y &= \left(X^T P^{-2} X\right)^{-1} X^T P^{-2} y \\
&= \left(X^T V^{-1} X\right)^{-1} X^T V^{-1} y,
\end{aligned}$$

since $\left(X^T V^{-1} X\right)^{-1}$ is non–singular. By Theorem 3.13 (Gauss–Markov Theorem), this is also the BLUE. $\qquad \square$

Note 4.35

By §3.3 the ordinary least squares estimator $\hat\beta = (X^T X)^{-1} X^T y$ is unbiased but by above is no longer the Best Linear Unbiased Estimator (BLUE).

Note 4.36

Theorem 4.34 is the key to a more general setting of mixed models (§9.1), where the BLUE is replaced by the best linear unbiased predictor (BLUP).

Note 4.37

In practice, if we do not assume that $V = I$ then the form that V should take instead is often unclear even if V is assumed diagonal as in *weighted least squares*. A pragmatic solution is first to perform the analysis of the data assuming $V = I$ and then to use the residuals of this model to provide an estimate $\hat{V}$ of V for use in a second stage analysis if this is deemed necessary. There appear to be no hard and fast ways of estimating V, and doing so in practice clearly depends on the precise experimental context. As an illustration, Draper and Smith (1998), Ch. 9, give an example of weighted regression assuming a quadratic relationship between a predictor and the squared residuals. See also Carroll and Ruppert (1988).

EXERCISES

4.1. *Polynomial regression* The data in Table 4.2 give the percentage of divorces caused by adultery per year of marriage. Investigate whether the rate of divorces caused by adultery is constant, and further whether or not a quadratic model in time is justified. Interpret your findings.

Year	1	2	3	4	5	6	7
%	3.51	9.50	8.91	9.35	8.18	6.43	5.31
Year	8	9	10	15	20	25	30
%	5.07	3.65	3.80	2.83	1.51	1.27	0.49

Table 4.2 Data for Exercise 4.1.

4.2. *Corner-point constraints and one-way ANOVA.* Formulate the regression model with k treatment groups as

$$
A = \begin{pmatrix}
1_{n_1} & 0_{n_2} & 0_{n_3} & \ldots & 0_{n_1} \\
1_{n_2} & 1_{n_2} & 0_{n_3} & \ldots & 0_{n_2} \\
1_{n_3} & 0_{n_2} & 1_{n_3} & \ldots & 0_{n_3} \\
\ldots & \ldots & \ldots & \ldots & \ldots \\
1_{n_k} & 0_{n_k} & 0_{n_k} & \ldots & 1_{n_k}
\end{pmatrix},
$$

$$
A^T A = \begin{pmatrix}
n_1 + n_2 + \ldots + n_k & n_2 & \ldots & n_k \\
n_2 & n_2 & \ldots & 0 \\
\ldots & \ldots & \ldots & \ldots \\
n_k & 0 & \ldots & n_k
\end{pmatrix},
$$

where n_j denotes the number of observations in treatment group j, 1_{n_j} is an associated n_j column vector of 1s and y_j denotes a column vector of observations corresponding to treatment group j.

(i) Show that

$$
A^T y = \begin{pmatrix}
n_1 \bar{y}_1 + n_2 \bar{y}_2 + \ldots + n_k \bar{y}_k \\
n_2 \bar{y}_2 \\
\ldots \\
n_k \bar{y}_k
\end{pmatrix}.
$$

(ii) In the case of two treatment groups calculate $\hat{\beta}$ and calculate the fitted values for an observation in each treatment group.

(iii) Show that

$$
M = (A^T A)^{-1} = \begin{pmatrix} \frac{1}{n_1} & \frac{-1}{n_1} & \cdots & \frac{-1}{n_1} \\ \frac{-1}{n_1} & \frac{1}{n_2} + \frac{1}{n_1} & \cdots & \frac{1}{n_1} \\ \cdots & \cdots & \cdots & \cdots \\ \frac{-1}{n_1} & \frac{1}{n_1} & \cdots & \frac{1}{n_k} + \frac{1}{n_1} \end{pmatrix}.
$$

Calculate $\hat{\beta}$, give the fitted values for an observation in treatment group j and interpret the results.

4.3. Fit the model in Example 2.9 using a regression approach.

4.4. Fit the model in Example 2.11 using a regression approach.

4.5. Define, $Y_0 \sim N(0, \sigma_0^2)$, $Y_i = Y_{i-1} + \epsilon_i$, where the ϵ_i are iid $N(0, \sigma^2)$. What is the joint distribution of
(i) Y_1, Y_2, Y_3,
(ii) $Y_1, \ldots, Y_n$?

4.6. Let $Y \sim N_3(\mu, \Sigma)$ with $\Sigma = \begin{pmatrix} 1 & a & 0 \\ a & 1 & b \\ 0 & b & 1 \end{pmatrix}$. Under what conditions are $Y_1 + Y_2 + Y_3$ and $Y_1 - Y_2 - Y_3$ independent?

4.7. *Mean-square prediction.* Let $Y \sim U(-a, b)$, $a, b > 0$, $X = Y^2$.
(i) Calculate $E(Y^n)$.
(ii) Find the best mean-square predictors of X given Y and of Y given X.
(iii) Find the best linear predictors of X given Y and of Y given X.

4.8. If the mean μ_0 in the multivariate normal distribution is known, show that the MLE of Σ is

$$
\hat{\Sigma} = \frac{1}{n} \sum_1^n (x_i - \mu_0)^T (x_i - \mu_0) = S + (\bar{x} - \mu_0)^T (\bar{x} - \mu_0).
$$

[Hint: Define the precision matrix $\Lambda = \Sigma^{-1}$ and use the differential rule $\partial/\partial A \ln |A| = (A^{-1})^T$.]

4.9. *Background results for Exercise 4.11.*
(i) Let $X \sim N(\mu, \Sigma)$. Show that

$$
f_X(x) \propto \exp\left\{ x^T A x + x^T b \right\},
$$

where $A = -\frac{1}{2} \Sigma^{-1}$ and $b = \Sigma^{-1} \mu$.
(ii) Let X and Y be two continuous random variables. Show that the conditional density $f_{X|Y}(x|y)$ can be expressed as $K f_{X,Y}(x, y)$ where K is constant with respect to x.

4.10. *Inverse of a partitioned matrix.* Show that the following formula holds for the inverse of a partitioned matrix:

$$
\begin{pmatrix} A & B \\ C & D \end{pmatrix}^{-1} = \begin{pmatrix} M & -MBD^{-1} \\ -D^{-1}CM & D^{-1} + D^{-1}CMBD^{-1} \end{pmatrix},
$$

where $M = (A - BD^{-1}C)^{-1}$. See e.g. Healy (1956), §3.4.

4.11. *Alternative derivation of conditional distributions in the multivariate normal family.* Let $X \sim N(\mu, \Sigma)$ and introduce the partition

$$
x = \begin{pmatrix} x_A \\ x_B \end{pmatrix}, \mu = \begin{pmatrix} \mu_A \\ \mu_B \end{pmatrix}, \Sigma = \begin{pmatrix} \Sigma_{AA} & \Sigma_{AB} \\ \Sigma_{BA} & \Sigma_{BB} \end{pmatrix}.
$$

Using Exercise 4.9 show that the conditional distribution of $x_A|x_B$ is multivariate normal with

$$
\begin{aligned}
\mu_{A|B} &= \mu_A + \Sigma_{AB}\Sigma_{BB}^{-1}(x_B - \mu_B), \\
\Sigma_{A|B} &= \Sigma_{AA} - \Sigma_{AB}\Sigma_{BB}^{-1}\Sigma_{BA}.
\end{aligned}
$$

Adding additional covariates and the Analysis of Covariance

5.1 Introducing further explanatory variables

Suppose that having fitted the regression model

$$\mathbf{y} = X\beta + \epsilon, \tag{M_0}$$

we wish to introduce q additional explanatory variables into our model. The augmented regression model, M_A, say becomes

$$\mathbf{y} = X\beta + Z\gamma + \epsilon. \tag{M_A}$$

We rewrite this as

$$\begin{aligned} \mathbf{y} = X\beta + Z\gamma + \epsilon &= (X, Z)\,(\beta, \gamma)^T + \epsilon, \\ &= W\delta + \epsilon, \end{aligned}$$

say, where

$$W := (X, Z), \quad \delta := \begin{pmatrix} \beta \\ \gamma \end{pmatrix}.$$

Here X is $n \times p$ and assumed to be of rank p, Z is $n \times q$ of rank q, and the columns of Z are linearly independent of the columns of X. This final assumption means that there is a sense in which the q additional explanatory variables are adding

N.H. Bingham and J.M. Fry, *Regression: Linear Models in Statistics*,
Springer Undergraduate Mathematics Series, DOI 10.1007/978-1-84882-969-5_5,
© Springer-Verlag London Limited 2010

genuinely new information to that already contained in the pre-existing X matrix. The least squares estimator $\hat{\delta}$ can be calculated directly, by solving the normal equations as discussed in Chapter 3, to give

$$\hat{\delta} = (W^T W)^{-1} W^T \mathbf{y}.$$

However, in terms of practical implementation, the amount of computation can be significantly reduced by using the estimate $\hat{\beta}$ obtained when fitting the model (M_0). We illustrate this method with an application to *Analysis of Covariance*, or *ANCOVA* for short. The results are also of interest as they motivate formal F-tests for comparison of nested models in Chapter 6.

Note 5.1

ANCOVA is an important subject in its own right and is presented here to illustrate further the elegance and generality of the general linear model as presented in Chapters 3 and 4. It allows one to combine, in a natural way, quantitative variables with qualitative variables as used in Analysis of Variance in Chapter 2. The subject was introduced by Fisher in 1932 (in §49.1 of the fourth and later editions of his book, Fisher (1958)). We proceed with the following lemma (where P is the projection or hat matrix, $P = X(X^T X)^{-1} X^T$ or $P = A(A^T A)^{-1} A^T = AC^{-1} A^T$ in our previous notation).

Lemma 5.2

If $R = I - P = I - X(X^T X)^{-1} X^T$, then $Z^T R Z$ is positive definite.

Proof

Suppose $\mathbf{x}^T Z^T R Z \mathbf{x} = 0$ for some vector $\mathbf{x}$. We have

$$\mathbf{x}^T Z^T R Z \mathbf{x} = \mathbf{x}^T Z^T R^T R Z \mathbf{x} = 0,$$

since R is idempotent from Lemma 3.18. It follows that $R Z \mathbf{x} = 0$, which we write as $Z \mathbf{x} = P Z \mathbf{x} = X \mathbf{y}$ say, for some vector $\mathbf{y}$. This implies $\mathbf{x} = 0$ as, by assumption, the columns of Z are linearly independent of the columns of X. Since $\mathbf{x} = 0$, it follows that $Z^T R Z$ is positive definite.   □

Theorem 5.3

Let $R_A = I - W(W^T W)^{-1} W^T$, $L = (X^T X)^{-1} X^T Z$ and

$$\hat{\delta} = \begin{pmatrix} \hat{\beta}_A \\ \hat{\gamma}_A \end{pmatrix}.$$

Then

(i) $\hat{\gamma}_A = (Z^T R Z)^{-1} Z^T R \mathbf{y}$,

(ii) $\hat{\beta}_A = (X^T X)^{-1} X^T (\mathbf{y} - Z \hat{\gamma}_A) = \hat{\beta} - L \hat{\gamma}_A$,

(iii) The sum of squares for error of the augmented model is given by

$$\mathbf{y}^T R_A \mathbf{y} = (\mathbf{y} - Z \hat{\gamma}_A)^T R (\mathbf{y} - Z \hat{\gamma}_A) = \mathbf{y}^T R \mathbf{y} - \hat{\gamma}_A Z^T R \mathbf{y}.$$

Proof

(i) We write the systematic component in the model equation (MA) as

$$\begin{aligned}
X\beta + Z\gamma &= X\beta + PZ\gamma + (I - P)Z\gamma, \\
&= X \left[\beta + (X^T X)^{-1} X^T Z\gamma \right] + RZ\gamma, \\
&= (X \quad RZ) \begin{pmatrix} \alpha \\ \gamma \end{pmatrix}, \\
&= V\lambda
\end{aligned}$$

say, where $\alpha = \beta + (X^T X)^{-1} X^T Z\gamma$. Suppose $V\lambda = 0$ for some λ. This gives $X\beta + Z\gamma = 0$, with both $\beta = \gamma = 0$ by linear independence of the columns of X and Z. Hence V has full rank $p + q$, since its null space is of dimension 0. From the definition $R = I - X(X^T X)^{-1} X^T$, one has $X^T R = RX = 0$. From Theorem 3.5, the normal equations can be solved to give

$$\begin{aligned}
\hat{\lambda} &= (V^T V)^{-1} V^T \mathbf{y}, \\
&= \begin{pmatrix} X^T X & X^T R Z \\ Z^T R X & Z^T R Z \end{pmatrix}^{-1} \begin{pmatrix} X^T \\ Z^T R \end{pmatrix} \mathbf{y}.
\end{aligned}$$

As $X^T R = RX = 0$, this product is

$$\begin{aligned}
\hat{\lambda} &= \begin{pmatrix} X^T X & 0 \\ 0 & Z^T R Z \end{pmatrix}^{-1} \begin{pmatrix} X^T \\ Z^T R \end{pmatrix} \mathbf{y} \\
&= \begin{pmatrix} (X^T X)^{-1} X^T \mathbf{y} \\ (Z^T R Z)^{-1} Z^T R \mathbf{y} \end{pmatrix}.
\end{aligned}$$

We can read off from the bottom row of this matrix

$$\hat{\gamma}_A = (Z^T R Z)^{-1} Z^T R \mathbf{y}.$$

(ii) From the top row of the same matrix,

$$\hat{\alpha} = (X^T X)^{-1} X^T \mathbf{y} = \hat{\beta},$$

since $\hat{\beta} = (X^T X)^{-1} X^T \mathbf{y}$. Since we defined $\alpha = \beta + (X^T X)^{-1} X^T Z \gamma$, it follows that our parameter estimates for the augmented model must satisfy

$$\hat{\alpha} = \hat{\beta}_A + \left(X^T X \right)^{-1} X^T Z \hat{\gamma}_A = \hat{\beta},$$

and the result follows.

(iii) We have that

$$
\begin{aligned}
R_A \mathbf{y} &= \mathbf{y} - X\hat{\beta}_A - Z\hat{\gamma}_A \\
&= \mathbf{y} - X \left(X^T X \right)^{-1} X^T (\mathbf{y} - Z\hat{\gamma}) - Z\hat{\gamma}_A \quad \text{(by (ii) and } (NE)) \\
&= \left(I - X \left(X^T X \right)^{-1} X^T \right) (\mathbf{y} - Z\hat{\gamma}_A) \\
&= R(\mathbf{y} - Z\hat{\gamma}_A) \\
&= R\mathbf{y} - RZ \left(Z^T R Z \right)^{-1} Z^T R\mathbf{y} \quad \text{(by (i)).}
\end{aligned}
$$

So by the above,

$$
\begin{aligned}
\mathbf{y}^T R_A \mathbf{y} &= \mathbf{y}^T RZ \left(Z^T R Z \right)^{-1} Z^T R\mathbf{y}, \\
&= \hat{\gamma}_A^T Z^T R\mathbf{y}.
\end{aligned}
$$

Since the matrices R_A and R are symmetric and idempotent (Lemma 3.18), the result can also be written as

$$
\begin{aligned}
\mathbf{y}^T R_A^T R_A \mathbf{y} &= \mathbf{y}^T R_A \mathbf{y} \\
&= (\mathbf{y} - Z\hat{\gamma}_A)^T R^T R(\mathbf{y} - Z\hat{\gamma}_A) \\
&= (\mathbf{y} - Z\hat{\gamma}_A)^T R(\mathbf{y} - Z\hat{\gamma}_A).
\end{aligned}
$$

$\square$

Sum of squares decomposition. We may rewrite (iii) as

$$SSE = SSE_A + \hat{\gamma}_A Z^T R\mathbf{y}.$$

That is, the sum of squares attributable to the new explanatory variables Z is

$$\hat{\gamma}_A Z^T R\mathbf{y}.$$

Linear hypothesis tests and an Analysis of Variance formulation based on a decomposition of sums of squares are discussed at length in Chapter 6. The result above gives a practical way of performing these tests for models which are constructed in a sequential manner. In particular, the result proves useful when fitting Analysis of Covariance models (§5.2–5.3).

One Extra Variable. The case with only one additional explanatory is worth special mention. In this case the matrix Z is simply a column vector, $x_{(p)}$ say. We have that $Z^T R Z = x_{(p)}^T R x_{(p)}$ is a scalar and the above formulae simplify to give

$$
\begin{aligned}
\hat{\gamma}_A &= \frac{x_{(p)}^T R \mathbf{y}}{x_{(p)}^T R x_{(p)}}, \\
\hat{\beta}_A &= \hat{\beta} - (X^T X)^{-1} X^T x_{(p)} \hat{\beta}_A, \\
\mathbf{y}^T R_A \mathbf{y} &= \mathbf{y}^T R \mathbf{y} - \hat{\gamma} x_{(p)}^T R \mathbf{y}.
\end{aligned}
$$

5.1.1 Orthogonal parameters

From Theorem 5.2(ii), the difference in our estimates of β in our two models, (M_0) and (M_A), is $L\hat{\gamma}_A$, where

$$
L := (X^T X)^{-1} X^T Z.
$$

Now $L = 0$ iff

$$
X^T Z = 0 \qquad\qquad (orth)
$$

(recall X is $n \times p$, Z is $n \times q$, so $X^T Z$ is $p \times q$, the matrix product being conformable). This is an orthogonality relation, not between vectors as usual but between matrices. When it holds, our estimates $\hat{\beta}$ and $\hat{\beta}_A$ of β in the original and augmented models (M_0) and (M_A) are the same. That is, if we are considering extending our model from (M_0) to (M_A), that is in extending our parameter from β to δ, we do not have to waste the work already done in estimating β, only to estimate the new parameter γ. This is useful and important conceptually and theoretically. It is also important computationally and in calculations done by hand, as was the case before the development of statistical packages for use on computers. As our interest is in the parameters (β, γ, δ) rather than the design matrices (X, Z, W), we view the orthogonality relation in terms of them, as follows:

Definition 5.4

In the above notation, the parameters β, γ are *orthogonal* (or β, γ are *orthogonal parameters*) if

$$
X^T Z = 0. \qquad\qquad (orth)
$$

Note 5.5

1. We have met such orthogonality before, in the context of polynomial regression (§4.1) and orthogonal polynomials (§4.1.2).

2. Even with computer packages, orthogonality is still an advantage from the point of view of numerical stability, as well as computational efficiency (this is why the default option in S-Plus uses orthogonal polynomials – see §4.1.3). Numerical stability is very important in regression, to combat one of the standing dangers – multicollinearity (see §7.4).

3. Orthogonal polynomials are useful in Statistics beyond regression. In statistical models with several parameters, it often happens that we are interested in some but not all of the parameters needed to specify the model. In this case, the (vector) parameter we are interested in – β, say – is (naturally) called the parameter of interest, or *interest parameter*, while the complementary parameter we are not interested in – γ, say – is called the *nuisance parameter*. The simplest classical case is the normal model $N(\mu, \sigma^2)$. If we are interested in the mean μ only, and not the variance σ^2, then σ is a nuisance parameter. The point of the Student t-statistic

$$t := \sqrt{n-1}(\overline{X} - \mu)/S \sim t(n-1)$$

familiar from one's first course in Statistics is that it *cancels out* σ:

$$\sqrt{n}(\overline{X} - \mu)/\sigma \sim N(0,1), \quad nS^2/\sigma^2 \sim \chi^2(n-1), \quad \overline{X} \text{ and } S \text{ independent.}$$

The tasks of estimating μ with σ known and with σ unknown are fundamentally different (and this is reflected in the difference between the normal and the t distributions).

Again, it may happen that with two parameters, θ_1 and θ_2 say, we have two statistics S_1 and S_2, such that while S_2 is uninformative about θ_1 on its own, (S_1, S_2) is more informative about θ_1 than S_1 alone is. One then says that the statistic S_2 is *ancillary* for inference about θ_1. Ancillarity (the concept is again due to Fisher) is best studied in conjunction with *sufficiency*, which we met briefly in §4.4.1. and §4.5.

With such issues in mind, one may seek to find the simplest, or most tractable, way to formulate the problem. It can be very helpful to *reparametrise*, so as to work with *orthogonal parameters*. The relevant theory here is due to Cox and Reid (1987) (D. R. Cox (1924–) and Nancy Reid (1952–)). Loosely speaking, orthogonal parameters allow one to separate a statistical model into its component parts.

5.2 ANCOVA

Recall that in regression (Chapters 1, 3, and 4) we have *continuous* (*quantitative*) variables, whilst in ANOVA (Chapter 2) we have *categorical* (qualitative) variables. For questions involving *both* qualitative *and* quantitative variables, we need to combine the methods of regression and ANOVA. This hybrid approach is Analysis of Covariance (ANCOVA).

Example 5.6

Suppose we want to compare two treatments A, B for reducing high blood pressure. Now blood pressure y is known to increase with age x (as the arteries deteriorate, by becoming less flexible, or partially blocked with fatty deposits, etc.). So we need to include *age* as a quantitative variable, called a *covariate* or *concomitant variable*, while we look at the treatments (qualitative variable), the variable of interest.

Suppose first that we inspect the data (EDA). See Figure 5.1, where x is age in years, y is blood pressure (in suitable units), the circles are those with treatment A and the triangles are those with treatment B.

This suggests the model

$$y_i = \begin{cases} \beta_{0A} + \beta_1 x_i + \epsilon_i & \text{for Treatment A;} \\ \beta_{0B} + \beta_1 x_i + \epsilon_i & \text{for Treatment B.} \end{cases}$$

This is the full model (of parallel-lines type in this example): there is a common slope, that is increase in age has the same effect for each treatment.

Here the *parameter of interest* is the *treatment effect*, or *treatment difference*, $\beta_{0A} - \beta_{0B}$, and the *hypothesis of interest* is that this is zero: $H_0 : \beta_{0A} = \beta_{0B}$.

Now see what happens if we ignore age as a covariate. In effect, this projects the plot above onto the y-axis. See Figure 5.2. The effect is much less clear!

Rewrite the model as ($\mu_i := Ey_i$; $E\epsilon_i = 0$ as usual)

$$\mu_i = \begin{cases} \beta_0 + \beta_1 x_i & \text{for Treatment A;} \\ \beta_0 + \beta_1 x_i + \beta_2 & \text{for Treatment B} \end{cases}$$

and test

$$H_0 : \beta_2 = 0.$$

The full model is: β_2 unrestricted.

The reduced model is: $\beta_2 = 0$.

Thus we are testing a *linear hypothesis* $\beta_2 = 0$ here.

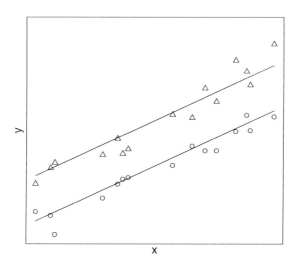

Figure 5.1 EDA plot suggests model with two different intercepts

We can put the quantitative variable x and the qualitative variable treatment on the same footing by introducing an *indicator* (or *Boolean*) variable,

$$z_i := \begin{cases} 0 & \text{if the } i\text{th patient has Treatment A,} \\ 1 & \text{if the } i\text{th patient has Treatment B.} \end{cases}$$

Then

– Full model: $\mu_i = \beta_0 + \beta_1 x_i + \beta_2 z$,

– Reduced model: $\mu_i = \beta_0 + \beta_1 x_i$,

– Hypothesis: $H_0 : \beta_2 = 0$.

As with regression and ANOVA, we might expect to test hypotheses using an F-test ('variance-ratio test'), with large values of an F-statistic significant against the null hypothesis. This happens with ANCOVA also; we come to the distribution theory later.

Interactions. The effect above is *additive* – one treatment simply *shifts* the regression line vertically relative to the other – see Figure 5.1. But things may

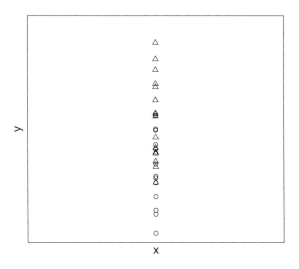

Figure 5.2 Ignorance of covariate blurs the ease of interpretation

be more complicated. For one of the treatments, say, there may be a decreasing treatment effect – the treatment effect may decrease with age, giving rise to non-parallel lines. The two lines may converge with age (when the treatment that seems better for younger patients begins to lose its advantage), may cross (when one treatment is better for younger patients, the other for older patients), or diverge with age (when the better treatment for younger patients looks better still for older ones). See Figure 5.3.

The full model now has four parameters (two general lines, so two slopes and two intercepts):

$$\mu_i = \beta_0 + \beta_1 x_i + \beta_2 z_i + \beta_3 z_i x_i \qquad \text{(general lines)},$$

the interaction term in β_3 giving rise to separate slopes.

The first thing to do is to test whether we need two separate slopes, by testing

$$H_0 : \beta_3 = 0.$$

If we do not, we simplify the model accordingly, back to

$$\mu_i = \beta_0 + \beta_1 x_i + \beta_2 z_i \qquad \text{(parallel lines)}.$$

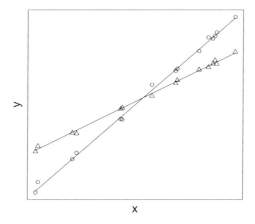

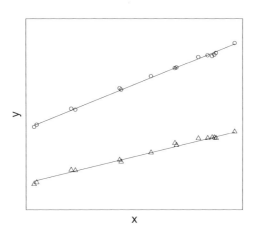

Figure 5.3 Top panel: Interaction term leads to convergence and then cross-over for increasing x. Bottom panel: Interaction term leads to divergence of treatment effects.

We can then test for treatment effect, by testing

$$H_0 : \beta_2 = 0.$$

If the treatment $(\beta_2 z)$ term is not significant, we can reduce again, to

$$\mu_i = \beta_0 + \beta_1 x_i \qquad \text{(common line)}.$$

We could, for completeness, then test for an age effect, by testing

$$H_0 : \beta_1 = 0$$

(though usually we would not do this – we know blood pressure *does* increase with age). The final, minimal model is

$$\mu_i = \beta_0.$$

These four models – with one, two, three and four parameters – are *nested models*. Each is successively a *sub-model* of the 'one above', with one more parameter. Equally, we have *nested hypotheses*

$$\beta_3 = 0,$$
$$\beta_2(= \beta_3) = 0,$$
$$\beta_1(= \beta_2 = \beta_3) = 0.$$

Note 5.7

In the medical context above, we are interested in *treatments* (which is the better?). But we are only able to test for a treatment effect if there is *no interaction*. Otherwise, it is not a question of the *better treatment*, but of which treatment is better *for whom*.

5.2.1 Nested Models

Update. Using a full model, we may wish to simplify it by deleting non-significant terms. Some computer packages allow one to do this by using a special command. In S-Plus/R® the relevant command is update. F-tests for nested models may simply be performed as follows:
```
m1.lm<-lm(y~x variables)
m2.lm<-update(a.lm, ~. -x variables to be deleted)
anova(m1.lm, m2.lm, test="F")
```
Note the syntax: to delete a term, use update and
$$, \ \sim . - \text{ "comma tilde dot minus"}.$$

Akaike Information Criterion (AIC). If there are p parameters in the model,

$$AIC := -2\text{log-likelihood} + 2(p+1)$$

(p parameters, plus one for σ^2, the unknown variance). We then choose between competing models by trying to *minimise* AIC. The AIC is a *penalised log-likelihood*, penalised by the number of parameters (H. Akaike (1927–) in 1974).

The situation is like that of polynomial regression (§4.1). Adding more parameters gives a better fit. But, the Principle of Parsimony tells us to use as few parameters as possible. AIC gives a sensible compromise between

bad fit, over-simplification, too few parameters, and

good fit, over-interpretation, too many parameters.

Step. One can test the various sub-models nested within the full model automatically in S-Plus, by using the command `step`. This uses AIC to drop non-significant terms (Principle of Parsimony: the fewer terms, the better). The idea is to start with the *full* model, and end up with the *minimal adequate* model.

Unfortunately, it matters in what *order* the regressors or factors are specified in our current model. This is particularly true in *ill-conditioned* situations (Chapter 7), where the problem is numerically unstable. This is usually caused by *multicollinearity* (some regressors being nearly linear combinations of others). We will discuss multicollinearity and associated problems in more detail in Chapter 7. F-tests for nested models and stepwise methods for model selection are further discussed in Chapter 6.

5.3 Examples

Example 5.8 (Photoperiod example revisited)

Here we suppose that the data in Exercises 2.4 and 2.9 can be laid out as in Table 5.1 – we assume we have quantitative rather than purely qualitative information about the length of time that plants are exposed to light. We demonstrate that Analysis of Covariance can lead to a flexible class of models by combining methods from earlier chapters on regression and Analysis of Variance.

The simplest model that we consider is *Growth~Genotype+Photoperiod*. This model has a different intercept for each different genotype. However, length of exposure to light is assumed to have the same effect on each plant irrespective of genotype. We can test for the significance of each term using an Analysis of Variance formulation analogous to the construction in Chapter 2. The sums-of-squares calculations are as follows. The total sum of squares and the genotype

Photoperiod	8h	12h	16h	24h
Genotype A	2	3	3	4
Genotype B	3	4	5	6
Genotype C	1	2	1	2
Genotype D	1	1	2	2
Genotype E	2	2	2	2
Genotype F	1	1	2	3

Table 5.1 Data for Example 5.8

sum of squares are calculated in exact accordance with the earlier analysis-of-variance calculations in Chapter 2:

$$SS = 175 - (1/24)57^2 = 39.625,$$
$$SSG = (1/4)(12^2 + 18^2 + 6^2 + 6^2 + 8^2 + 7^2) - (1/24)57^2 = 27.875.$$

As before we have 23 total degrees of freedom and 5 degrees of freedom for genotype. In Chapter 1 we saw that the sum of squares explained by regression is given by

$$SSR := \sum_i (\hat{y}_i - \overline{y})^2 = \frac{S_{xy}^2}{S_{xx}}.$$

Since photoperiod is now assumed to be a quantitative variable, we have only one degree of freedom in the ANOVA table. The sum-of-squares calculation for photoperiod becomes $77^2/840 = 7.058$. As before, the residual sum of squares is calculated by subtraction.

In the notation of Theorem 5.3 we find that

$$
Z = \begin{pmatrix}
0 & 0 & 0 & 0 & 0 \\
0 & 0 & 0 & 0 & 0 \\
0 & 0 & 0 & 0 & 0 \\
0 & 0 & 0 & 0 & 0 \\
1 & 0 & 0 & 0 & 0 \\
1 & 0 & 0 & 0 & 0 \\
1 & 0 & 0 & 0 & 0 \\
1 & 0 & 0 & 0 & 0 \\
0 & 1 & 0 & 0 & 0 \\
0 & 1 & 0 & 0 & 0 \\
0 & 1 & 0 & 0 & 0 \\
0 & 1 & 0 & 0 & 0 \\
0 & 0 & 1 & 0 & 0 \\
0 & 0 & 1 & 0 & 0 \\
0 & 0 & 1 & 0 & 0 \\
0 & 0 & 1 & 0 & 0 \\
0 & 0 & 0 & 1 & 0 \\
0 & 0 & 0 & 1 & 0 \\
0 & 0 & 0 & 1 & 0 \\
0 & 0 & 0 & 1 & 0 \\
0 & 0 & 0 & 0 & 1 \\
0 & 0 & 0 & 0 & 1 \\
0 & 0 & 0 & 0 & 1 \\
0 & 0 & 0 & 0 & 1
\end{pmatrix}, \quad X = \begin{pmatrix}
1 & 8 \\
1 & 12 \\
1 & 16 \\
1 & 24 \\
1 & 8 \\
1 & 12 \\
1 & 16 \\
1 & 24 \\
1 & 8 \\
1 & 12 \\
1 & 16 \\
1 & 24 \\
1 & 8 \\
1 & 12 \\
1 & 16 \\
1 & 24 \\
1 & 8 \\
1 & 12 \\
1 & 16 \\
1 & 24 \\
1 & 8 \\
1 & 12 \\
1 & 16 \\
1 & 24
\end{pmatrix}.
$$

Using $\hat{\gamma}_A = (Z^T R Z)^{-1} Z^T R Y$ gives

$$
\hat{\gamma}_A = \begin{pmatrix}
1.5 \\
-1.5 \\
-1.5 \\
-1 \\
-1.25
\end{pmatrix}.
$$

The regression sum of squares for genotype can then be calculated as $\hat{\gamma}_A Z^T R Y = 27.875$ and we obtain, by subtraction, the resulting ANOVA table in Table 5.2. All terms for photoperiod and genotype are significant and we appear to need a different intercept term for each genotype.

A second model that we consider is *Photoperiod~Genotype*Photoperiod*. This model is a more complicated extension of the first, allowing for the possibility of different intercepts *and* different slopes, dependent on genotype. As before, the degrees of freedom multiply to give five degrees of freedom for this

Source	df	Sum of Squares	Mean Square	F	p
Photoperiod	1	7.058	7.058	25.576	0.000
Genotype	5	27.875	5.575	20.201	0.000
Residual	17	4.692	0.276		
Total	23	39.625			

Table 5.2 ANOVA table for different intercepts model

interaction term. The sum-of-squares term of the Genotype:Photoperiod interaction term can be calculated as follows. In the notation of Theorem 5.3, we now have

$$
Z = \begin{pmatrix}
0 & 0 & 0 & 0 & 0 \\
0 & 0 & 0 & 0 & 0 \\
0 & 0 & 0 & 0 & 0 \\
0 & 0 & 0 & 0 & 0 \\
8 & 0 & 0 & 0 & 0 \\
12 & 0 & 0 & 0 & 0 \\
16 & 0 & 0 & 0 & 0 \\
24 & 0 & 0 & 0 & 0 \\
0 & 8 & 0 & 0 & 0 \\
0 & 12 & 0 & 0 & 0 \\
0 & 16 & 0 & 0 & 0 \\
0 & 24 & 0 & 0 & 0 \\
0 & 0 & 8 & 0 & 0 \\
0 & 0 & 12 & 0 & 0 \\
0 & 0 & 16 & 0 & 0 \\
0 & 0 & 24 & 0 & 0 \\
0 & 0 & 0 & 8 & 0 \\
0 & 0 & 0 & 12 & 0 \\
0 & 0 & 0 & 16 & 0 \\
0 & 0 & 0 & 24 & 0 \\
0 & 0 & 0 & 0 & 8 \\
0 & 0 & 0 & 0 & 12 \\
0 & 0 & 0 & 0 & 16 \\
0 & 0 & 0 & 0 & 24
\end{pmatrix}, \quad
X = \begin{pmatrix}
1 & 8 & 0 & 0 & 0 & 0 & 0 \\
1 & 12 & 0 & 0 & 0 & 0 & 0 \\
1 & 16 & 0 & 0 & 0 & 0 & 0 \\
1 & 24 & 0 & 0 & 0 & 0 & 0 \\
1 & 8 & 1 & 0 & 0 & 0 & 0 \\
1 & 12 & 1 & 0 & 0 & 0 & 0 \\
1 & 16 & 1 & 0 & 0 & 0 & 0 \\
1 & 24 & 1 & 0 & 0 & 0 & 0 \\
1 & 8 & 0 & 1 & 0 & 0 & 0 \\
1 & 12 & 0 & 1 & 0 & 0 & 0 \\
1 & 16 & 0 & 1 & 0 & 0 & 0 \\
1 & 24 & 0 & 1 & 0 & 0 & 0 \\
1 & 8 & 0 & 0 & 1 & 0 & 0 \\
1 & 12 & 0 & 0 & 1 & 0 & 0 \\
1 & 16 & 0 & 0 & 1 & 0 & 0 \\
1 & 24 & 0 & 0 & 1 & 0 & 0 \\
1 & 8 & 0 & 0 & 0 & 1 & 0 \\
1 & 12 & 0 & 0 & 0 & 1 & 0 \\
1 & 16 & 0 & 0 & 0 & 1 & 0 \\
1 & 24 & 0 & 0 & 0 & 1 & 0 \\
1 & 8 & 0 & 0 & 0 & 0 & 1 \\
1 & 12 & 0 & 0 & 0 & 0 & 1 \\
1 & 16 & 0 & 0 & 0 & 0 & 1 \\
1 & 24 & 0 & 0 & 0 & 0 & 1
\end{pmatrix}.
$$

$\hat{\gamma}_A = (Z^T R Z)^{-1} Z^T R Y$ gives

$$\hat{\gamma}_A = \begin{pmatrix} 0.071 \\ -0.071 \\ -0.043 \\ -0.114 \\ 0.021 \end{pmatrix}.$$

The sum of squares for the Genotype:Photoperiod term (Gen:Phot.) can then be calculated as $\hat{\gamma}_A Z^T R Y = 3.149$ and we obtain the ANOVA table shown in Table 5.3. We see that the Genotype:Photoperiod interaction term is significant and the model with different slopes and different intercepts offers an improvement over the simpler model with just one slope but different intercepts.

Source	df	Sum of Squares	Mean Square	F	p
Photoperiod	1	7.058	7.058	54.898	0.000
Genotype	5		5.575	43.361	0.000
Gen:Phot.	5	3.149	0.630	4.898	0.011
(Different slopes)					
Residual	12	1.543	0.129		
Total	23	39.625			

Table 5.3 ANOVA table for model with different intercepts and different slopes

Example 5.9 (Exercise 1.6 revisited)

We saw a covert Analysis of Covariance example as early as the Exercises at the end of Chapter 1, in the half-marathon times in Table 1.2. The first model we consider is a model with different intercepts. The sum of squares for age is $114.795^2/747.5 = 17.629$. Fitting the model suggested in part (ii) of Exercise 1.6 gives a residual sum of squares of 43.679. The total sum of squares is $SS = 136.114$. Substituting gives a sum of squares of $136.114 - 43.679 - 17.629 = 74.805$ for club status. This result can alternatively be obtained as follows. We have that

$$Z = (0, 0, 0, 1, 1, 1, 1, 1, 1, 1, 1, 1, 1, 1)^T,$$

$$X = \begin{pmatrix} 1 & 1 & 1 & 1 & 1 & 1 & 1 & 1 & 1 & 1 & 1 & 1 & 1 & 1 \\ 42 & 43 & 44 & 46 & 48 & 49 & 50 & 51 & 57 & 59 & 60 & 61 & 62 & 63 \end{pmatrix}^T.$$

We have that $\hat{\gamma}_A = (Z^T R Z)^{-1} Z^T R Y = -7.673$ and the sum of squares for club status can be calculated as $\hat{\gamma}_A(Z^T R Y) = (-7.673)(-9.749) = 74.805$.

The ANOVA table obtained is shown in Table 5.4. The term for club status is significant, but the age term is borderline insignificant. The calculations for the model with two different slopes according to club status is left as an exercise (see Exercise 5.1).

Source	df	Sum of squares	Mean Square	F	p
Age	1	17.629	17.629	4.440	0.059
Club membership	1	74.805	74.805	18.839	0.001
Residual	11	43.679	3.971		
Total	13	136.114			

Table 5.4 ANOVA table for different intercepts model

EXERCISES

5.1. Produce the ANOVA table for the model with different slopes for the data in Example 5.9.

5.2. In the notation of Theorem 5.3 show that

$$\operatorname{var}\left(\hat{\delta}_A\right) = \begin{pmatrix} (X^T X)^{-1} - LML^T & +LM \\ -ML^T & M \end{pmatrix},$$

where $M = (Z^T RZ)^{-1}$.

5.3. Suppose $Y_1, \ldots, Y_n$ are iid $N(\alpha, \sigma^2)$.
(i) Find the least-squares estimate of α.
(ii) Use Theorem 5.3 to estimate the augmented model

$$Y_i = \alpha + \beta x_i + \epsilon_i,$$

and verify the formulae for the estimates of the simple linear regression model in Chapter 1.

5.4. Repeat the analysis in Chapter 5.3 in S-Plus/R® using the commands update and anova.

5.5. The data in Table 5.5 come from an experiment measuring enzymatic reaction rates for treated (State=1) and untreated (State=0) cells exposed to different concentrations of substrate. Fit an Analysis of Covariance model to this data and interpret your findings.

State=0		State=1	
Concentration	Rate	Concentration	Rate
0.02	67	0.02	76
0.02	51	0.02	47
0.06	84	0.06	97
0.06	86	0.06	107
0.11	98	0.11	123
0.11	115	0.11	139
0.22	131	0.22	159
0.22	124	0.22	152
0.56	144	0.56	191
0.56	158	0.56	201
1.10	160	1.10	207
		1.10	200

Table 5.5 Data for Exercise 5.5

5.6. *ANCOVA on the log-scale.* Plot the data in Exercise 5.5. Does the assumption of a linear relationship appear reasonable? Log-transform both the independent variable and the response and try again. (This suggests a power-law relationship; these are extremely prevalent in the physical sciences.) Fit an Analysis of Covariance model and write out your final fitted model for the experimental rate of reaction.

5.7. The data in Table 5.6 is telephone usage (in 1000s) in various parts of the world. Fit an Analysis of Covariance model to the logged data, with time as an explanatory variable, using a different intercept term for each region. Test this model against the model with a different intercept *and* a different slope for each country.

	N. Am.	Europe	Asia	S. Am.	Oceania	Africa	Mid Am.
51	45939	21574	2876	1815	1646	89	555
56	60423	29990	4708	2568	2366	1411	733
57	64721	32510	5230	2695	2526	1546	773
58	68484	35218	6662	2845	2691	1663	836
59	71799	37598	6856	3000	2868	1769	911
60	76036	40341	8220	3145	3054	1905	1008
61	79831	43173	9053	3338	3224	2005	1076

Table 5.6 Data for Exercise 5.7

5.8. *Quadratic Analysis of Covariance model.* Suppose we have one explanatory variable X but that the data can also be split into two categories as denoted by a dummy variable Z. Write

$$Y = \beta_0 + \beta_1 X + \beta_2 X^2 + \gamma_0 Z + \gamma_1 ZX + \gamma_2 ZX^2 + \epsilon.$$

In addition to the possibility of different intercepts and different slopes this model allows for additional curvature, which can take different forms in each category. Suppose the first k observations are from the first category $(Z = 0)$ and the remaining $n - k$ are from the second category $(Z = 1)$.
(i) Write down the X matrix for this model.
Suggest appropriate F-tests to test:
(ii) The need for both quadratic terms,
(iii) The hypothesis $\gamma_2 = 0$ assuming $\beta_2 \neq 0$.

5.9. *Probability plots/normal probability plots.* Given an ordered sample x_i, an approximate test of normality can be defined by equating the theoretical and empirical cumulative distribution functions (CDFs):

$$\frac{i}{n} = \Phi\left(\frac{x_i - \mu}{\sigma}\right),$$

where $\Phi(\cdot)$ is the standard normal CDF. In practice, to avoid boundary effects, the approximate relation

$$\frac{i - \frac{1}{2}}{n} = \Phi\left(\frac{x_i - \mu}{\sigma}\right)$$

is often used (a 'continuity correction'; cf. Sheppard's correction, Kendall and Stuart (1977) §3.18–26).
(i) Use this approximate relation to derive a linear relationship and suggest a suitable graphical test of normality.
(ii) The following data represent a simulated sample of size 20 from $N(0, 1)$. Do these values seem reasonable using the above?

-2.501, -1.602, -1.178, -0.797, -0.698, -0.428, -0.156, -0.076, -0.032, 0.214, 0.290, 0.389, 0.469, 0.507, 0.644, 0.697, 0.820, 1.056, 1.145, 2.744

[Hint: In S-Plus/R® you may find the commands `ppoints` and `qqnorm` helpful.]
(iii) A random variable on $[0, L]$ has a power-law distribution if it has probability density $f(x) = ax^b$. Find the value of a and derive

an approximate goodness-of-fit test for this distribution by equating theoretical and empirical CDFs.

5.10. *Segmented/piecewise linear models.* Suppose we have the following data:

$$x \;=\; (1, 2, 3, 4, 5, 6, 7, 8, 9),$$
$$y \;=\; (1.8, 4.3, 5.6, 8.2, 9.1, 10.7, 11.5, 12.2, 14.0).$$

Suppose it is known that a change-point occurs at $x = 5$, so that observations 1–4 lie on one straight line and observations 5–9 lie on another.

(i) Using dummy variables express this model as a linear model. Write down the X matrix. Fit this model and interpret the fitted parameters.

(ii) Assume that the location of the change-point is unknown and can occur at each of $x = \{4, 5, 6, 7\}$. Which choice of change-point offers the best fit to data?

(iii) Show that for a linear regression model the maximised likelihood function can be written as $\propto SSE$. Hence, show that AIC is equivalent to the penalty function

$$n \ln(SSE) + 2p.$$

Hence, compare the best fitting change-point model with linear and quadratic regression models with no change-point.

6
Linear Hypotheses

6.1 Minimisation Under Constraints

We have seen several examples of hypotheses on models encountered so far. For example, in dealing with polynomial regression §4.1 we met, when dealing with a polynomial model of degree k, the hypothesis that the degree was at most $k - 1$ (that is, that the leading coefficient was zero). In Chapter 5, we encountered nested models, for example two general lines, including two parallel lines. We then met the hypothesis that the slopes were in fact equal (and so the lines were parallel). We can also conduct a statistical check of structural constraints (for instance, that the angles of a triangle sum to two right-angles – see Exercise 6.5).

We thus need to formulate a general framework for hypotheses of this kind, and for testing them. Since the whole thrust of the subject of regression is linearity, it is to be expected that our attention focuses on linear hypotheses.

The important quantities are the parameters β_i, $i = 1, \ldots, p$. Thus one expects to be testing hypotheses which impose linear constraints on these parameters. We shall be able to test k such constraints, where $k \leq p$. Assembling these into matrix form, we shall test a *linear hypothesis* (with respect to the parameters) of the matrix form

$$B\beta = c. \qquad\qquad (hyp)$$

Here B is a $k \times p$ matrix, β is the $p \times 1$ vector of parameters, and c is a $k \times 1$ vector of constants. We assume that matrix B has full rank: if not, there are linear

N.H. Bingham and J.M. Fry, *Regression: Linear Models in Statistics*,
Springer Undergraduate Mathematics Series, DOI 10.1007/978-1-84882-969-5_6,
© Springer-Verlag London Limited 2010

dependencies between rows of B; we then avoid redundancy by eliminating dependent rows, until remaining rows are linearly independent and B has full rank. Since $k \leq p$, we thus have that B has rank k.

We now seek to minimise the total sum of squares SS, with respect to variation of the parameters β, subject to the constraint (hyp). Now by (SSD) of §3.4,

$$SS = SSR + SSE.$$

Here SSE is a statistic, and can be calculated from the data y; it does not involve the unknown parameters β. Thus our task is actually to

$$\text{minimise} \qquad SSR = (\hat{\beta} - \beta)^T C (\hat{\beta} - \beta) \qquad \text{under} \qquad B\beta = c.$$

This *constrained minimisation* problem is solved by introducing *Lagrange multipliers*, $\lambda_1, \ldots, \lambda_k$, one for each component of the constraint equation (hyp). We solve instead the *unconstrained mimimisation problem*

$$\min \qquad \frac{1}{2} SSR + \lambda^T (B\beta - c),$$

where λ is the k-vector with ith component λ_i. Readers unfamiliar with Lagrange multipliers are advised to take the method on trust for the moment: we will soon produce our minimising value, and demonstrate that it does indeed achieve the minimum – or see e.g. Dineen (2001), Ch. 3 or Ostaszewski (1990), §15.6. (See also Exercises 6.4–6.6.) That is, we solve

$$\min \qquad \frac{1}{2} \sum\sum_{i,j=1}^{p} c_{ij} \left(\hat{\beta}_i - \beta_i \right) \left(\hat{\beta}_j - \beta_j \right) + \sum_{i=1}^{k} \lambda_j \left(\sum_{j=1}^{p} b_{ij}\beta_j - c_i \right).$$

For each $r = 1, \ldots, k$, we differentiate partially with respect to β_r and equate the result to zero. The double sum gives two terms, one with $i = r$ and one with $j = r$; as $C = (c_{ij})$ is symmetric, we obtain

$$-\sum_j c_{jr} \left(\hat{\beta}_j - \beta_j \right) + \sum_i \lambda_i b_{ir} = 0.$$

The terms above are the rth elements of the vectors $-C(\hat{\beta} - \beta)$ and $B^T \lambda$. So we may write this system of equations in matrix form as

$$B^T \lambda = C \left(\hat{\beta} - \beta \right). \qquad (a)$$

Now C is positive definite, so C^{-1} exists. Pre-multiply by BC^{-1} (B is $k \times p$, C^{-1} is $p \times p$):

$$BC^{-1}B^T \lambda = B \left(\hat{\beta} - \beta \right) = B\hat{\beta} - c,$$

by (hyp). Since C^{-1} is positive definite ($p \times p$) and B is full rank ($k \times p$), $BC^{-1}B^T$ is positive definite ($k \times k$). So we may solve for λ, obtaining

$$\lambda = \left(BC^{-1}B^T \right)^{-1} (B\hat{\beta} - c). \qquad (b)$$

We may now solve (a) and (b) for β, obtaining

$$\beta = \hat{\beta} - C^{-1}B^T \left(BC^{-1}B^T\right)^{-1} \left(B\hat{\beta} - c\right).$$

This is the required minimising value under (hyp), which we write as $\beta^\dagger$:

$$\beta^\dagger = \hat{\beta} - C^{-1}B^T \left(BC^{-1}B^T\right)^{-1} \left(B\hat{\beta} - c\right). \qquad (c)$$

In $SSR = (\hat{\beta} - \beta)^T C(\hat{\beta} - \beta)$, replace $\hat{\beta} - \beta$ by $(\hat{\beta} - \beta^\dagger) + (\beta^\dagger - \beta)$. This gives two squared terms, and a cross term,

$$2(\beta^\dagger - \beta)^T C(\hat{\beta} - \beta^\dagger),$$

which by (a) is

$$2(\beta^\dagger - \beta)^T B\lambda.$$

But $B\beta = c$ and $B\beta^\dagger = c$, by (hyp). So $B(\beta^\dagger - \beta) = 0$, $(\beta^\dagger - \beta)^T B = 0$, and the cross term is zero. So

$$SSR = (\hat{\beta} - \beta)^T C(\hat{\beta} - \beta) = (\hat{\beta} - \beta^\dagger)^T C(\hat{\beta} - \beta^\dagger) + (\beta^\dagger - \beta)^T C(\beta^\dagger - \beta). \quad (d)$$

The second term on the right is non-negative, and is zero only for $\beta = \beta^\dagger$, giving

Theorem 6.1

Under the linear constraint (hyp), the value

$$\beta^\dagger = \hat{\beta} - C^{-1}B^T(BC^{-1}B^T)^{-1}(B\hat{\beta} - c)$$

is the unique minimising value of the quadratic form SSR in β.
(i) The unique minimum of SS under (hyp) is

$$SS^* = SSR + (\hat{\beta} - \beta^\dagger)^T C(\hat{\beta} - \beta^\dagger).$$

Multiplying (c) by B confirms that $B\beta^\dagger = c$ – that is, that $\beta^\dagger$ does satisfy (hyp). Now (d) shows directly that $\beta^\dagger$ is indeed the minimising value of SS and so of SS. Thus those unfamiliar with Lagrange multipliers may see directly from (d) that the result of the theorem is true.

Proposition 6.2

$E(SS^*) = (n - p + k)\sigma^2.$

Proof

The matrix B is $k \times p$ ($k \le p$), and has full rank k. So some $k \times k$ sub–matrix of B is non-singular. We can if necessary relabel columns so that the first k columns form this non-singular $k \times k$ sub–matrix. We can then solve the linear system of equations

$$B\beta = c$$

to find $\beta_1, \ldots, \beta_k$ – in terms of the remaining parameters $\beta_{k+1}, \ldots, \beta_{k+p}$. We can then express SS as a function of these $p - k$ parameters, and solve by ordinary least squares. This is then *unconstrained* least squares with $p - k$ parameters. We can then proceed as in Chapter 3 but with $p - k$ in place of p, obtaining $E(SS^*) = (n - p + k)\sigma^2$.   □

6.2 Sum-of-Squares Decomposition and F-Test

Definition 6.3

The *sum of squares for the linear hypothesis*, SSH, is the difference between the constrained minimum SS^* and the unconstrained minimum SSE of SS. Thus

$$SSH := SS^* - SSE = (\hat{\beta} - \beta^\dagger)^T C(\hat{\beta} - \beta^\dagger).$$

We proceed to find its distribution. As usual, we reduce the distribution theory to matrix algebra, using symmetric projections.

　　Now

$$\hat{\beta} - \beta^\dagger = C^{-1}B^T \left(BC^{-1}B^T\right)^{-1} \left(B\hat{\beta} - c\right),$$

by (i) of the Theorem above. So

$$B\hat{\beta} - c = B\left(\hat{\beta} - \beta\right) + (B\beta - c) = B\left(\hat{\beta} - \beta\right),$$

under the constraint (*hyp*). But

$$
\begin{aligned}
\hat{\beta} - \beta &= C^{-1}A^T y - \beta \\
&= C^{-1}A^T y - C^{-1}A^T A\beta \\
&= C^{-1}A^T(y - A\beta).
\end{aligned}
$$

Combining,

$$\hat{\beta} - \beta^\dagger = C^{-1}B^T \left(BC^{-1}B^T\right)^{-1} BC^{-1}A^T(y - A\beta),$$

so we see that

$$\left(\hat{\beta} - \beta^{\dagger}\right)^T C = (y - A\beta)^T AC^{-1} B^T \left(BC^{-1} B^T\right) BC^{-1} C$$
$$= (y - A\beta)^T AC^{-1} B^T \left(BC^{-1} B^T\right) B.$$

Substituting these two expressions into the definition of SSH above, we see that SSH is

$$(y - A\beta)^T AC^{-1} B^T \left(BC^{-1} B^T\right)^{-1} B.C^{-1} B^T \left(BC^{-1} B^T\right)^{-1} BC^{-1} A^T (y - A\beta),$$

which simplifies, giving

$$SSH = (y - A\beta)^T D(y - A\beta),$$

say, where

$$D := AC^{-1} B^T \left(BC^{-1} B^T\right)^{-1} BC^{-1} A^T.$$

Now matrix D is *symmetric*, and

$$D^2 = AC^{-1} B^T \left(BC^{-1} B^T\right)^{-1} BC^{-1} A^T.AC^{-1} B^T \left(BC^{-1} B^T\right)^{-1} BC^{-1} A^T$$

which simplifies to

$$D^2 = AC^{-1} B^T \left(BC^{-1} B^T\right)^{-1} BC^{-1} A^T$$
$$= D,$$

so D is also *idempotent*. So its rank is its trace, and D is a symmetric projection.

By the definition of SS^*, we have the sum-of-squares decomposition

$$SS^* := SSE + SSH.$$

Take expectations:
$$E(SS^*) = E(SSE) + E(SSH).$$

But
$$E(SSE) = (n - p)\sigma^2,$$

by §3.4, and
$$E(SS^*) = (n - p + k)\sigma^2,$$

by Proposition 6.2 above. Combining,

$$E(SSH) = k\sigma^2.$$

Since SSH is a quadratic form in normal variates with matrix D, a symmetric projection, this shows as in §3.5.1, that D has rank k:

$$rank(D) = \text{trace}(D) = k,$$

the number of (scalar) constraints imposed by the (matrix) constraint (*hyp*).

Theorem 6.4 (Sum of Squares for Hypothesis, SSH)

(i) In the sum-of-squares decomposition

$$SS^* := SSE + SSH,$$

the terms on the right are independent.

(ii) The three quadratic forms are chi-square distributed, with

$$SS^*/\sigma^2 \sim \chi^2(n - p + k), \qquad SSE/\sigma^2 \sim \chi^2(n - p), \qquad SSH/\sigma^2 \sim \chi^2(k).$$

Proof

Since the ranks $n - p$ and k of the matrices of the quadratic forms on the right sum to the rank $n - p + k$ of that on the left, and we already know that quadratic forms in normal variates are chi-square distributed, the independence follows from Chi-Square Decomposition, §3.5.   □

We are now ready to formulate a test of our linear hypothesis (hyp). This use of Fisher's F distribution to test a general linear hypothesis is due to S. Kołodziejcyzk (d. 1939) in 1935.

Theorem 6.5 (Kołodziejcyzk's Theorem)

We can test our linear hypothesis (hyp) by using the F-statistic

$$F := \frac{SSH/k}{SSE/(n - p)},$$

with large values of F evidence against (hyp). Thus at significance level α, we use critical region

$$F > F_\alpha(k, n - p),$$

the upper α-point of the Fisher F-distribution $F(k, n - p)$.

Proof

By the result above and the definition of the Fisher F-distribution as the ratio of independent chi-square variates divided by their degrees of freedom, our F-statistic has distribution $F(k, n - p)$. It remains to show that *large* values of F are evidence *against* (hyp) – that is, that a one-tailed test is appropriate.

Write

$$w = B\beta - c.$$

Thus $w = 0$ iff the linear hypothesis (hyp) is true; w is non-random, so constant (though unknown, as it involves the unknown parameters β). Now

$$B\hat{\beta} - c = B\left(\hat{\beta} - \beta\right) + (B\beta - c) = B\left(\hat{\beta} - \beta\right) + w.$$

Here $\hat{\beta} - \beta = C^{-1}A^T(y - A\beta)$ has mean zero and covariance matrix $\sigma^2 C^{-1}$ (Proposition 4.4). So $B\hat{\beta} - c$ and $B(\hat{\beta} - \beta)$ have covariance matrix $\sigma^2 BC^{-1}B^T$; $B(\hat{\beta} - \beta)$ has mean zero (as $\hat{\beta}^*$ is unbiased), and $B\beta - c$ has mean w. Now by Theorem 6.1,

$$
\begin{aligned}
SSH &= (\hat{\beta} - \beta^\dagger)^T C(\hat{\beta} - \beta^\dagger) \\
&= [C^{-1}B^T\left(BC^{-1}B^T\right)^{-1}(B\hat{\beta} - c)]^T C[C^{-1}B^T(BC^{-1}B^T)^{-1}(B\hat{\beta} - c)].
\end{aligned}
$$

This is a quadratic form in $B\hat{\beta} - c$ (mean w, covariance matrix $\sigma^2 BC^{-1}B^T$) with matrix

$$(BC^{-1}B^T)^{-1}.BC^{-1}.C.C^{-1}B^T(BC^{-1}B^T)^{-1} = (BC^{-1}B^T)^{-1}.$$

So by the Trace Formula (Prop. 3.22),

$$E(SSH) = \text{trace}[(BC^{-1}B^T)^{-1}.\sigma^2 BC^{-1}B^T] + w^T(BC^{-1}B^T)^{-1}w.$$

The trace term is $\sigma^2\text{trace}(I_k)$ (B is $k \times p$, C^{-1} is $p \times p$, B^T is $p \times k$), or $\sigma^2 k$, giving

$$E(SSH) = \sigma^2 k + w^T(BC^{-1}B^T)^{-1}w.$$

Since C is positive definite, so is C^{-1}, and as B has full rank, so is $(BC^{-1}B^T)^{-1}$. The second term on the right is thus non-negative, and positive unless $w = 0$; that is, unless the linear hypothesis (hyp) is true. Thus large values of $E(SSH)$, so of SSH, so of $F := (SSH/k)/(SSE/(n-p))$, are associated with violation of (hyp). That is, a one-tailed test, rejecting (hyp) if F is too big, is appropriate. $\square$

Note 6.6

The argument above makes no mention of distribution theory. Thus it holds also in the more general situation where we do not assume *normally distributed* errors, only uncorrelated errors with the same variance. A one-tailed F-test is indicated there too. However, the difficulty comes when choosing the critical region – the cut-off level above which we will reject the null hypothesis – the linear hypothesis (hyp). With normal errors, we know that the F-statistic has the F-distribution $F(k, n - p)$, and we can find the cut-off level $F_\alpha(k, n - p)$ using the significance level α and tables of the F-distribution. Without the assumption of normal errors, we do not know the distribution of the F-statistic – so although we still know that large values are evidence against (*hyp*), we lack a yardstick to tell us 'how big is too big'. In practice, we would probably still use tables of the F-distribution, 'by default'. This raises questions of how close to normality our error distribution is, and how sensitive to departures from normality the distribution of the F-statistic is – that is, how *robust* our procedure is against departures from normality. We leave such robustness questions to the next chapter, but note in passing that Robust Statistics is an important subject in its own right, on which many books have been written; see e.g. Huber (1981).

Note 6.7

To implement this procedure, we need to proceed as follows.

(i) Perform the regression analysis in the 'big' model, Model 1 say, obtaining our SSE, SSE_1 say.

(ii) Perform the regression analysis in the 'little' model, Model 2 say, obtaining similarly SSE_2.

(iii) The big model gives a better fit than the little model; the difference in fit is $SSH := SSE_2 - SSE_1$.

(iv) We normalise the difference in fit SSH by the number k of degrees of freedom by which they differ, obtaining SSH/k.

(v) This is the numerator of our F-statistic. The denominator is SSE_1 divided by its df.

This procedure can easily be implemented by hand – it is after all little more than two regression analyses. Being both so important and so straightforward, it has been packaged, and is automated in most of the major statistical packages.

In S-Plus/R®, for example, this procedure is embedded in the software used whenever we compare two nested models, and in particular in the automated procedures `update` and `step` of §5.2. As we shall see in §6.3 the theory motivates a host of sequential methods to automatically select from the range of possible models.

Example 6.8 (Brownlee's stack loss data)

This data set is famous in statistics for the number of times it has been analysed. The data in Table 6.1 relate stack loss – a measure of inefficiency – to a series of observations. Exploratory data analysis suggests close relationships between Stack Loss and Air Flow and between Water Temperature and Stack Loss.

We wish to test whether or not Acid Concentration can be removed from the model. This becomes a test of the hypothesis $\alpha_3 = 0$ in the model

$$Y = \alpha_0 + \alpha_1 X_1 + \alpha_2 X_2 + \alpha_3 X_3 + \epsilon.$$

Air Flow X_1			Water Temp X_2			Acid Conc. X_3			Stack Loss Y		
80	62	50	27	24	18	89	93	89	42	20	8
80	58	50	27	23	18	88	87	86	37	15	7
75	58	50	25	18	19	90	80	72	37	14	8
62	58	50	24	18	19	87	89	79	28	14	8
62	58	50	22	17	20	87	88	80	18	13	9
62	58	56	23	18	20	87	82	82	18	11	15
62	58	70	24	19	20	93	93	91	19	12	15

Table 6.1 Data for Example 6.8

Fitting the model with all three explanatory variables gives a residual sum of squares of 178.83 on 17 df The model with acid concentration excluded has a residual sum of squares of 188.795 on 16 df Our F-statistic becomes

$$F = \left(\frac{188.795 - 178.83}{1} \right) \left(\frac{16}{188.795} \right) = 0.85.$$

Testing against $F_{1,16}$ gives a p-value of 0.372. Thus, we accept the null hypothesis and conclude that Acid Concentration can be excluded from the model.

6.3 Applications: Sequential Methods

6.3.1 Forward selection

We start with the model containing the constant term. We consider all the explanatory variables in turn, choosing the variable for which SSH is largest. The procedure is repeated for $p = 2, 3, \ldots$, selecting at each stage the variable not currently included in the model with largest F statistic. The procedure terminates when either all variables are included in the model or the maximum F value fails to exceed some threshold F_{IN}.

Example 6.9

We illustrate forward selection by returning to the data in Example 6.8.

Step 1
We compute SSE(Air Flow) $= 319.116$, SSE(Water Temperature) $= 483.151$, SSE(Acid concentration) $= 1738.442$. Air flow is the candidate for entry into the model. $F = 104.201$ against $F_{1,19}$ to give $p = 0.000$ so air flow enters the model.

Step 2
The computations give SSE(Air Flow+Water Temperature) $= 188.795$ and SSE(Air Flow+Acid Concentration) $= 309.1376$. Thus, water temperature becomes our candidate for entry into the model. We obtain that $F = 12.425$ and testing against $F_{1,18}$ gives $p = 0.002$ so water temperature enters the model.

Step 3
The F-test of Example 6.8 shows that acid concentration does not enter the model.

6.3.2 Backward selection

Backward selection is an alternative to forward selection. We start using the full model using all p variables (recall $p << n$) and compute the F-statistic with $k = 1$ for each of the p-variables in turn. We eliminate the variable having smallest F-statistic from the model, provided F is less than some threshold F_{OUT}. The procedure is continued until either all the variables are excluded from the model or the smallest F fails to become less than F_{OUT}. When performing forward or backward selection the thresholds F_{IN} and F_{OUT} may change as the algorithms proceed. The most obvious approach is to choose an appropriate formal significance level, e.g. $p = 0.05$, and set the thresholds according to the critical values of the corresponding F-test.

Example 6.10

We illustrate backward selection by returning to the example.

Step 1
The F-test of Example 6.8 excludes acid concentration from the model.

Step 2

The calculations show that SSE(Air Flow +Water Temperature) $= 188.795$, SSE(Air Flow) $= 319.116$, SSE(Water Temperature) $= 483.151$. Thus water temperature becomes our candidate for exclusion. The resulting F-test is the same as in Step 2 of Example 6.9, and we see that no further terms can be excluded from the model.

6.3.3 Stepwise regression

In forward selection, once a variable is included in the model it is not removed. Similarly, in backward selection once a variable is excluded it is never reintroduced. The two algorithms may also give very different results when applied to the same data set. *Stepwise regression* aims to resolve these issues by combining forward selection and backward selection.

The algorithm starts with the simple model consisting solely of a constant term. The first step is a forward selection stage, followed by a backward selection step. The algorithm then alternates between forward and backward selection steps until no further variables are introduced at the forward selection stage. It is shown in Seber and Lee (2003) Ch. 12 that if $F_{OUT} \leq F_{IN}$ then the algorithm must eventually terminate.

Example 6.11 (Example 6.8 re-visited)

The forward selection steps see first Air Flow and then Water Temperature enter the model. Example 6.10 then shows that neither of these variables can be excluded at the backward selection phase. Example 6.8 then shows that Acid Concentration cannot enter the model in the final forward selection phase.

Note 6.12

Some additional discussion of stepwise methods can be found in Seber and Lee (2003), Ch. 12. The S-Plus/R® command `step` uses a variant of the above method based on AIC (§5.2.1), which works both with Linear Models (Chapters 1–7) and Generalised Linear Models (Chapter 8). The command `step` can also be used to perform forward and backward selection by specifying `direction`.

EXERCISES

6.1. Fit regression models to predict fuel consumption for the data set shown in Table 6.2 using
(i) Forward selection
(ii) Backward selection
(iii) Stepwise regression.
T is a qualitative variable taking the value 1 specifying a manual rather than an automatic gearbox. G denotes the number of gears, C denotes the number of carburettors. RAR is the rear-axle ratio, 1/4M t is the time taken to complete a quarter of a mile circuit. Cyls. gives the number of cylinders and Disp. is the car's displacement. (This is a classical data set extracted from the 1974 Motor Trend US magazine, and available as part of the mtcars dataset in R®.)

6.2. Show that the first step in forward selection is equivalent to choosing the variable most highly correlated with the response.

6.3. *All-subsets regression.*
(i) Suppose that we have p non-trivial explanatory variables and we always include a constant term. Show that the number of possible models to consider in all–subsets regression is $2^p - 1$.
(ii) How many possible models are suggested in Exercise 6.1?
(iii) Suppose it is feasible to fit no more than 100 regression models. How large does p have to be in order for all-subsets regression to become infeasible?

6.4. *Lagrange multipliers method.* Using the Lagrange multipliers method maximise $f(x, y) := xy$ subject to the constraint $x^2 + 8y^2 = 4$. [Hint: Set $L = xy + \lambda(x^2 + 8y^2 - 4)$, where λ is the Lagrange multiplier, and differentiate with respect to x and y. The resulting solution for λ transforms the constrained problem into an unconstrained problem.]

6.5. *Angles in a triangle.* A surveyor measures three angles of a triangle, α, β, γ ($\alpha + \beta + \gamma = \pi$). Given one measurement of each of these angles, find the constrained least–squares solution to this problem by using Lagrange multipliers.

6.6. *Angles in a cyclic quadrilateral.* A surveyor measures four angles α, β, γ, δ which are known to satisfy the constraint $\alpha + \beta + \gamma + \delta = 2\pi$. If there is one observation for each of these angles Y_1, Y_2, Y_3, Y_4 say, find the constrained least–squares solution to this problem using Lagrange multipliers.

Mpg	Cyls.	Disp.	Hp	RAR	Weight	1/4M t	v/s	T.	G.	C.
21.0	6	160.0	110	3.90	2.620	16.46	0	1	4	4
21.0	6	160.0	110	3.90	2.875	17.02	0	1	4	4
22.8	4	108.0	93	3.85	2.320	18.61	1	1	4	1
21.4	6	258.0	110	3.08	3.215	19.44	1	0	3	1
18.7	8	360.0	175	3.15	3.440	17.02	0	0	3	2
18.1	6	225.0	105	2.76	3.460	20.22	1	0	3	1
14.3	8	360.0	245	3.21	3.570	15.84	0	0	3	4
24.4	4	146.7	62	3.69	3.190	20.00	1	0	4	2
22.8	4	140.8	95	3.92	3.150	22.90	1	0	4	2
19.2	6	167.6	123	3.92	3.440	18.30	1	0	4	4
17.8	6	167.6	123	3.92	3.440	18.90	1	0	4	4
16.4	8	275.8	180	3.07	4.070	17.40	0	0	3	3
17.3	8	275.8	180	3.07	3.730	17.60	0	0	3	3
15.2	8	275.8	180	3.07	3.780	18.00	0	0	3	3
10.4	8	472.0	205	2.93	5.250	17.98	0	0	3	4
10.4	8	460.0	215	3.00	5.424	17.82	0	0	3	4
4.7	8	440.0	230	3.23	5.345	17.42	0	0	3	4
32.4	4	78.7	66	4.08	2.200	19.47	1	1	4	1
30.4	4	75.7	52	4.93	1.615	18.52	1	1	4	2
33.9	4	71.1	65	4.22	1.835	19.90	1	1	4	1
21.5	4	120.1	97	3.70	2.465	20.01	1	0	3	1
15.5	8	318.0	150	2.76	3.520	16.87	0	0	3	2
15.2	8	304.0	150	3.15	3.435	17.30	0	0	3	2
13.3	8	350.0	245	3.73	3.840	15.41	0	0	3	4
19.2	8	400.0	175	3.08	3.845	17.05	0	0	3	2
27.3	4	79.0	66	4.08	1.935	18.90	1	1	4	1
26.0	4	120.3	91	4.43	2.140	16.70	0	1	5	2
30.4	4	95.1	113	3.77	1.513	16.90	1	1	5	2
15.8	8	351.0	264	4.22	3.170	14.50	0	1	5	4
19.7	6	145.0	175	3.62	2.770	15.50	0	1	5	6
15.0	8	301.0	335	3.54	3.570	14.60	0	1	5	8
21.4	4	121.0	109	4.11	2.780	18.60	1	1	4	2

Table 6.2 Data for Exercise 6.1

6.7. Show that the regression treatment of one-way ANOVA and the F-test for linear hypotheses returns the original F-test in Theorem 2.8.

6.8. Use a regression formulation and a suitable F-test to test the hypothesis of no differences between treatments in Example 2.9.

6.9. Repeat Exercise 6.1, this time treating the 1/4M time as the dependant variable.

6.10. *Mixtures.* Often chemical experiments involve mixtures of ingredients. This introduces a constraint into the problem, typically of the form

$$x_1 + x_2 + \ldots + x_p = 1.$$

Suppose $x_1, \ldots, x_p$ are from a mixture experiment and satisfy the above constraint.
(i) Reformulate the full main effects model

$$y_i = \beta_0 + \beta_1 x_{1,i} + \ldots + \beta_p x_{p,i} + \epsilon_i,$$

using this constraint.
(ii) Suppose $p = 3$. The usual full second-order model is

$$
\begin{aligned}
y &= \beta_0 + \beta_1 x_1 + \beta_2 x_2 + \beta_3 x_3 + \beta_{11} x_1^2 + \beta_{12} x_1 x_2 + \beta_{13} x_1 x_3 \\
&+ \beta_{22} x_2^2 + \beta_{23} x_2 x_3 + \beta_{33} x_3^2 + \epsilon.
\end{aligned}
$$

Using your answer to (i) suggest a possible way to estimate this model. What is the general solution to this problem for $p \neq 3$?

6.11. *Testing linear hypotheses.*
(i) Test for the need to use a quadratic model in order to describe the following mixture experiment. $x_1 = (1, 0, 0, 0.5, 0.5, 0, 0.2, 0.3)$, $x_2 = (0, 1, 0, 0.5, 0, 0.5, 0.6, 0.5)$, $x_3 = (0, 0, 1, 0, 0.5, 0.5, 0.2, 0.2)$, $y = (40.9, 25.5, 28.6, 31.1, 24.9, 29.1, 27.0, 28.4)$.
(ii) Suppose we have the following data $x_1 = (-1, -1, 0, 1, 1)$, $x_2 = (-1, 0, 0, 0, 1)$, $y = (7.2, 8.1, 9.8, 12.3, 12.9)$. Fit the model $y = \beta_0 + \beta_1 x_1 + \beta_2 x_2 + \epsilon$. Test the hypothesis that $\beta_1 = 2\beta_2$. Explain how this constrained model may be fitted using simple linear regression.

7

Model Checking and Transformation of Data

7.1 Deviations from Standard Assumptions

In the above, we have assumed several things:

(i) the mean $\mu = Ey$ is a linear function of the regressors, or of the parameters;

(ii) the errors are additive;

(iii) the errors are independent;

(iv) the errors are normally distributed (Gaussian);

(v) the errors have equal variance.

Any or all of these assumptions may be inadequate. We turn now to a discussion of how to assess the adequacy of our assumptions, and to what we can do when they are inadequate.

Residual Plots. We saw in §3.6 that the residuals e_i and fitted values y_i^* are independent. So a residual plot of e_i against y_i^* should not show any particular pattern. If it *does*, then this suggests that the model is inadequate.

Scatter Plots. Always begin with EDA. With one regressor, we look at the scatter plot of y_i against x_i. With more than one regressor, one can look at all scatter plots of pairs of variables. In S-Plus, this can be done by using the command `pairs`. For details, see for example the S-Plus Help facility, or Crawley (2002), Ch. 24 (especially p. 432–3).

N.H. Bingham and J.M. Fry, *Regression: Linear Models in Statistics,*
Springer Undergraduate Mathematics Series, DOI 10.1007/978-1-84882-969-5_7,
© Springer-Verlag London Limited 2010

With two regressors, we have a data cloud in three dimensions. This is a highly typical situation: real life is lived in three spatial dimensions, but we represent it – on paper, or on computer screens – in two dimensions. The mathematics needed for this – the mathematics of computer graphics, or of virtual reality – is based on projective geometry. In S-Plus, the command `brush` allows one, in effect, to 'pick up the data cloud and rotate it' (see the S-Plus Help facility, or Venables and Ripley (2002), for details). This may well reveal important structural features of our data. For example, if the data appears round from one direction, but elliptical from another, this tells one something valuable about its distribution, and may suggest some appropriate transformation of the data.

In higher dimensions, we lose the spatial intuition that comes naturally to us in three dimensions. This is a pity, but is unavoidable: many practical situations involve more than two regressors, and so more than three dimensions. One can still use `pairs` to look at two-dimensional scatter plots, but there are many more of these to look at, and combining these different pieces of visual information is not easy.

In higher dimensions, the technique of Projection Pursuit gives a systematic way of searching for adequate low-dimensional descriptions of the data.

Non-constant Variance. In Figure 7.2 the points 'fan out' towards the right, suggesting that the variance increases with the mean. One possibility is to use *weighted regression* (§4.7). Another possibility is to *transform* the data (see below and Draper and Smith (1998) Ch. 13 for further details).

Unaccounted-for Structure. If there is visible structure present, e.g. curvature, in the residual plot, this suggests that the model is not correct. We should return to the original scatter plot of y against x and reinspect. One possibility is to consider adding an extra term or terms to the model – for example, to try a quadratic rather than a linear fit, etc.

Outliers. These are unusual observations that do not conform to the pattern of the rest of the data. They are always worth *checking* (e.g., has the value been entered correctly, has a digit been mis-transcribed, has a decimal point been slipped, etc.?)

Such outliers may be unreliable, and distort the reliable data. If so, we can *trim* the data to remove them. On the other hand, such points, if genuine, may be highly informative.

The subject of how to get protection against such data contamination by removing aberrant data points is called Robust Statistics (touched on in §5.3). In particular, we can use Robust Regression.

Example 7.1 (Median v Mean)

As a measure of location (or central tendency), using medians rather than means gives us some protection against aberrant data points. Indeed, medians can withstand gross data contamination – up to half the data wrong – without failing completely (up to half the data can go off to infinity without dragging the median off to infinity with them). We say that the median has *breakdown point* $1/2$, while the mean has breakdown point zero.

Detecting outliers via residual analysis. Residual analysis can be useful in gauging the extent to which individual observations may be expected to deviate from the underlying fitted model. As above, large residuals may point to problems with the original data. Alternatively they may indicate that a better model is needed, and suggest ways in which this may be achieved. The raw residuals are given by

$$e_i = y_i - x_i \hat{\beta}.$$

Scaled residuals are defined as

$$e_i^* = \frac{e_i}{\sqrt{m_{ii}}},$$

where the m_{ii} are the diagonal elements of the matrix M, where $M = I - P = I - X(X^T X)^{-1} X^T$. Under this construction the scaled residuals should now have equal variances (see Theorem 3.30). Scaled residuals can be further modified to define *standardised* or *internally studentised residuals* defined as

$$s_i = \frac{e_i^*}{\hat{\sigma}}.$$

The distribution of the internally studentised residuals is approximately t_{n-p}. However, the result is not exact since the numerator and denominator are not independent. There is one further type of residual commonly used: the *standardised deletion* or *externally studentised residual*. Suppose we wish to test the influence that observation i has on a fitted regression equation. Deleting observation i and refitting we obtain a *deletion residual*

$$e_{-i} = y_i - x_i^T \hat{\beta}_{-i},$$

where $\hat{\beta}_{-i}$ is the estimate obtained *excluding* observation i. Working as above we can define a *standardised* deletion residual s_{-i}. It can be shown, see e.g. Seber and Lee (2003) Ch. 10, that

$$s_{-i} = \frac{s_i \sqrt{n-p-1}}{\sqrt{n-p-s_i^2}}.$$

Further, if the model is correctly defined, these externally studentised residuals have an *exact* t_{n-p-1} distribution. Residual plots can be generated automatically in S-Plus/R® using the command `plot`. In R® this produces a plot of residuals against fitted values, a normal probability plot of standardised residuals (the relevant command here is `qqnorm`), a plot of the square root of the absolute standardised residuals against fitted values, and a plot of standardised residuals versus leverage with control limits indicating critical values for Cook's distances. (See below for further details.)

Influential Data Points. A point has high *leverage* if omitting it causes a big change in the fit. For example, with one regressor x, an x_i far from $\bar{x}$ with an atypical y_i will have high leverage. The leverage of observation i is given by h_{ii} – the diagonal elements of the hat matrix H or projection matrix P. In R® the leverages can be retrieved using the command `hat`. As an illustration we consider an admittedly contrived example in Huber (1981) and also cited in Atkinson (1985). Data consist of $x = -4, -3, -2, -1, 0, 10$, $y = 2.48, 0.73, -0.04, -1.44, -1.32, 0.00$ and the effect of including or excluding the apparent outlier at $x = 10$ has a dramatic impact upon the line of best fit (see Figure 7.1).

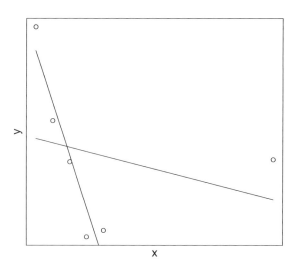

Figure 7.1 Effect of influential observation on line of best fit

Cook's distance. The Cook's distance D_i of observation i combines leverage and residuals – as can be seen from the definition (here $H = (h_{ij}) = P$)

$$D_i = \frac{s_i^2 h_{ii}}{p(1 - h_{ii})}.$$

Large values of Cook's distance occur if an observation is *both* outlying (large s_i) with high leverage (large h_{ii}). Plots of Cook's distance can be obtained as part of the output automatically generated in S-Plus/R® using the command `plot`. It can be shown that

$$D_i = \frac{\left(\hat{\beta} - \hat{\beta}_{-i}\right)^T X^T X \left(\hat{\beta} - \hat{\beta}_{-i}\right)}{p\hat{\sigma}^2},$$

where $\hat{\beta}_{-i}$ is the parameter estimate $\hat{\beta}$ obtained when the ith observation is *excluded*. Thus D_i does indeed serve as a measure of the *influence* of observation i. It provides an appropriate measure of the 'distance' from $\hat{\beta}$ to $\hat{\beta}_{-i}$.

Note 7.2

1. For further background on Cook's distance and related matters, we refer to Cook and Weisberg (1982).

2. This 'leave one out' idea is often useful in statistics. It leads to the method of *cross-validation* (CV).

Bias and Mallows's C_p statistic.   Suppose we fit the model

$$\mathbf{y} = X_1\beta_1 + \epsilon.$$

This leads to the least-squares estimate $\hat{\beta}_1 = (X_1^T X_1)^{-1} X_1^T \mathbf{y}$. If our postulated model is correct then this estimate is unbiased (§3.3). Suppose however that the true underlying relationship is

$$\mathbf{y} = X_1\beta_1 + X_2\beta_2 + \epsilon.$$

Our least-squares estimate $\hat{\beta}_1$ now has expected value $\beta_1 + (X_1^T X_1)^{-1} X_1^T X_2\beta_2$. Omitting X_2 leads to a bias of $(X_1^T X_1)^{-1} X_1^T X_2\beta_2$. Note that this is 0 if $X_1^T X_2 = 0$, the orthogonality relation we met in §5.1.1 on orthogonal parameters.

Mallows's C_p statistic is defined as

$$C_p = \frac{SSE}{s^2} - (n - 2p),$$

where p is the number of model parameters and s^2 is an estimate of σ^2 obtained from a subjective choice of full model. We consider sub-models of the full model. If a model is approximately correct

$$E(C_p) \approx \frac{(n-p)\sigma^2}{\sigma^2} - (n - 2p) = p.$$

If the model is incorrectly specified it is assumed $E(SSE) > \sigma^2$ and $E(C_p) > p$. Models can be compared using this method by plotting C_p against p. Suitable candidate models should lie close to the line $C_p = p$. Note, however that by definition $C_p = p$ for the full model.

Non-additive or non-Gaussian errors. These may be handled using Generalised Linear Models (see Chapter 8). Generalised Linear Models can be fitted in S-Plus and R® using the command `glm`. For background and details, see McCullagh and Nelder (1989).

Correlated Errors. These are always very dangerous in Statistics! *Independent* errors tend to *cancel*. This is the substance of the Law of Large Numbers (LLN), that says

$$\bar{x} \to Ex \qquad (n \to \infty)$$

– sample means tend to population means as sample size increases. Similarly for sample variances and other sample quantities. This is basically why Statistics works. One does not even need to have *independent* errors: weakly dependent errors (which may be defined precisely, in a variety of ways) exhibit similar cancellation behaviour. By contrast, *strongly dependent* errors need *not* cancel. Here, increasing the sample size merely replicates existing readings, and if these are way off this does not help us (as in Note 1.3).

Correlated errors may have some special structure – e.g., in time or in space. Accordingly, one would then have to use special methods to reflect this – Time Series or Spatial Statistics; see Chapter 9. Correlated errors may be detected using the Durbin–Watson test or, more crudely, using a runs test (see Draper and Smith (1998), Ch. 7).

7.2 Transformation of Data

If the residual plot 'funnels out' one may try a transformation of data, such as $y \mapsto \log y$ or $y \mapsto \sqrt{y}$ (see Figure 7.2).

If on the other hand the residual plot 'funnels in' one may instead try $y \mapsto y^2$, etc (see Figure 7.3).

Is there a general procedure? One such approach was provided in a famous paper Box and Cox (1964). Box and Cox proposed a one-parameter family of

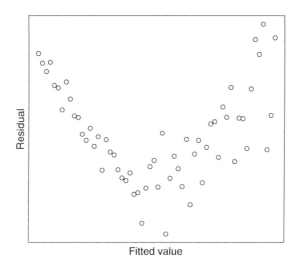

Figure 7.2 Plot showing 'funnelling out' of residuals

power transformations that included a *logarithmic* transformation as a special case. With λ as parameter, this is

$$y \mapsto \begin{cases} (y^\lambda - 1)/\lambda & \text{if} \quad \lambda \neq 0, \\ \log y & \text{if} \quad \lambda = 0. \end{cases}$$

Note that this is an indeterminate form at $\lambda = 0$, but since

$$\frac{y^\lambda - 1}{\lambda} = \frac{e^{\lambda \log y} - 1}{\lambda},$$

$$\frac{d}{d\lambda}\left(e^{\lambda \log y} - 1\right) = \log y . e^{\lambda \log y} = \log y \qquad \text{if } \lambda = 0,$$

L'Hospital's Rule gives

$$(y^\lambda - 1)/\lambda \to \log y \qquad (\lambda \to 0).$$

So we may *define* $(y^\lambda - 1)/\lambda$ as $\log y$ for $\lambda = 0$, to include $\lambda = 0$ with $\lambda \neq 0$ above.

One may – indeed, should – proceed *adaptively* by allowing the *data* to suggest which value of λ might be suitable. This is done in S-Plus by the command boxcox.

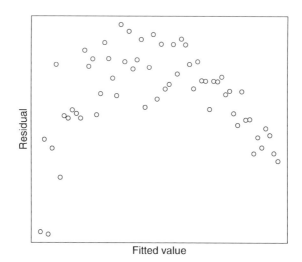

Figure 7.3 Plot showing 'funnelling in' of residuals

Example 7.3 (Timber Example)

The value of timber yielded by a tree is the response variable. This is measured only when the tree is cut down and sawn up. To help the forestry worker decide which trees to fell, the predictor variables used are girth ('circumference' – though the tree trunks are not perfect circles) and height. These can be easily measured without interfering with the tree – girth by use of a tape measure (at some fixed height above the ground), height by use of a surveying instrument and trigonometry.

Venables and Ripley (2002) contains a data library MASS, which includes a data set timber:

```
attach(timber)
names(timber)
[1] "volume" "girth" "height"
boxcox(volume) ~ (girth + height)
```

Dimensional Analysis. The data-driven choice of Box–Cox parameter λ seems to be close to $1/3$. This is predictable on dimensional grounds: volume is in cubic metres, girth and height in metres (or centimetres). It thus always pays to be aware of *units*.

There is a whole subject of *Dimensional Analysis* devoted to such things (see e.g. Focken (1953)). A background in Physics is valuable here.

7.3 Variance-Stabilising Transformations

In the exploratory data analysis (EDA), the scatter plot may suggest that the variance is not constant throughout the range of values of the predictor variable(s). But, the theory of the Linear Model *assumes* constant variance. Where this standing assumption seems to be violated, we may seek a systematic way to *stabilise* the variance – to make it constant (or roughly so), as the theory requires.

If the response variable is y, we do this by seeking a suitable function g (sufficiently smooth – say, twice continuously differentiable), and then *transforming* our data by

$$y \mapsto g(y).$$

Suppose y has mean μ:

$$Ey = \mu.$$

Taylor expand $g(y)$ about $y = \mu$:

$$g(y) = g(\mu) + (y - \mu)g'(\mu) + \frac{1}{2}(y - \mu)^2 g''(\mu) + \ldots$$

Suppose the bulk of the response values y are fairly closely bunched around the mean μ. Then, approximately, we can treat $y - \mu$ as small; then $(y - \mu)^2$ is negligible (at least to a first approximation, which is all we are attempting here). Then

$$g(y) \sim g(\mu) + (y - \mu)g'(\mu).$$

Take expectations: as $Ey = \mu$, the linear term goes out, giving $Eg(y) \sim g(\mu)$. So

$$g(y) - g(\mu) \sim g(y) - Eg(y) \sim g'(\mu)(y - \mu).$$

Square both sides:

$$[g(y) - g(\mu)]^2 \sim [g'(\mu)]^2 (y - \mu)^2.$$

Take expectations: as $Ey = \mu$ and $Eg(y) \sim g(\mu)$, this says

$$\mathrm{var}(g(y)) \sim [g'(\mu)]^2 \mathrm{var}(y).$$

Regression. So if
$$E(y_i|x_i) = \mu_i, \qquad \text{var}(y_i|x_i) = \sigma_i^2,$$
we use EDA to try to find some link between the means μ_i and the variances σ_i^2. Suppose we try $\sigma_i^2 = H(\mu_i)$, or
$$\sigma^2 = H(\mu).$$
Then by above,
$$\text{var}(g(y)) \sim [g'(\mu)]^2\sigma^2 = [g'(\mu)]^2 H(\mu).$$
We want *constant variance*, c^2 say. So we want
$$[g'(\mu)]^2 H(\mu) = c^2, \qquad g'(\mu) = \frac{c}{\sqrt{H(\mu)}}, \qquad g(y) = c\int \frac{dy}{\sqrt{H(y)}}.$$

Note 7.4

The idea of variance-stabilising transformations (like so much else in Statistics!) goes back to Fisher. He found the density of the sample correlation coefficient r^2 in the bivariate normal distribution – a complicated function involving the population correlation coefficient ρ^2, simplifying somewhat in the case $\rho = 0$ (see e.g. Kendall and Stuart (1977), §16.27, 28). But Fisher's z transformation of 1921 (Kendall and Stuart (1977), §16.33)
$$r = \tanh z, \qquad z = \frac{1}{2}\log\left(\frac{1+r}{1-r}\right), \qquad \rho = \tanh \zeta, \qquad \zeta = \frac{1}{2}\log\left(\frac{1+\rho}{1-\rho}\right)$$
gives z approximately normal, with variance almost independent of ρ:
$$z \sim N(0, 1/(n-1)).$$

Taylor's Power Law. The following empirical law was proposed by R. L. Taylor in 1961 (Taylor (1961)):
log variance against log mean is roughly *linear* with slope γ between 1 and 2.
   Both these extreme cases can occur. An example of slope 1 is the Poisson distribution, where the mean and the variance are the same. An example of slope 2 occurs with a Gamma-distributed error structure, important in Generalised Linear Models (Chapter 8).
   With $H(\mu) = \mu^\gamma$ above, this gives variance
$$v = \sigma^2 = H(\mu) = \mu^\gamma.$$
Transform to
$$g(y) = c\int \frac{dy}{\sqrt{H(y)}} = c\int \frac{dy}{y^{\frac{1}{2}\gamma}} = c\left(y^{1-\frac{1}{2}\gamma} - y_0^{1-\frac{1}{2}\gamma}\right).$$

This is of Box–Cox type, with

$$\lambda = 1 - \frac{1}{2}\gamma.$$

Taylor's suggested range $1 \leq \gamma \leq 2$ gives

$$0 \leq 1 - \frac{1}{2}\gamma \leq \frac{1}{2}.$$

Note that this range includes the logarithmic transformation (Box–Cox, $\lambda = 0$), and the cube–root transformation ($\lambda = 1/3$) in the timber example. Partly for dimensional reasons as above, common choices for λ include $\lambda = -1/2, 0, 1/3, 1/2, (1), 3/2$ (if $\lambda = 1$ we do not need to transform). An empirical choice of λ (e.g. by Box–Cox as above) close to one of these may suggest choosing λ as this value, and/or a theoretical examination with dimensional considerations in mind.

Delta Method. A similar method applies to *reparametrisation*. Suppose we choose a parameter θ. If the true value is θ_0 and the maximum-likelihood estimator is $\hat{\theta}$, then under suitable regularity conditions a central limit theorem (CLT) will hold:

$$\sqrt{n}\left(\hat{\theta} - \theta_0\right)/\sigma \to N(0,1) \qquad (n \to \infty).$$

Now suppose that one wishes to change parameter, and work instead with ϕ, where

$$\phi := g(\theta).$$

Then the same method (Taylor expansion about the mean) enables one to transfer this CLT for our estimate of θ to a CLT for our estimate of ϕ:

$$\sqrt{n}\left(\hat{\phi} - \phi_0\right)/\left(g'(\theta_0)\,\sigma\right) \to N(0,1) \qquad (n \to \infty).$$

Example 7.5 (Variance and standard deviation)

It is convenient to be able to change at will from using variance σ^2 as a parameter to using standard deviation σ. Mathematically the change is trivial, and it is also trivial computationally (given a calculator). Using the delta-method, it is statistically straightforward to transfer the results of a maximum-likelihood estimation from one to the other.

7.4 Multicollinearity

Recall the distribution theory of the bivariate normal distribution (§1.5). If we are regressing y on x, *but y is* (exactly) a *linear* function of x, then $\rho = \pm 1$, the bivariate normal density does not exist, and the *two*-dimensional setting is wrong – the situation is really *one*-dimensional. Similar remarks apply for the multivariate normal distribution (§4.3). When we assume the covariance matrix Σ is non-singular, the density exists and is given by Edgeworth's Theorem; when Σ is singular, the density does not exist. The situation is similar again in the context of Multiple Regression in Chapter 3. There, we assumed that the design matrix A ($n \times p$, with $n >> p$) has *full rank p*. A will have defective rank ($< p$) if there are linear relationships between regressors. In all these cases, we have a general situation which is non-degenerate, but which contains a special situation which is degenerate. The right way to handle this is to identify the degeneracy and its cause. By reformulating the problem in a suitably lower dimension, we can change the situation which is degenerate in the higher-dimensional setting into one which is non-degenerate if handled in its natural dimension. To summarise: to escape degeneracy, one needs to identify the linear dependence relationship which causes it. One can then eliminate dependent variables, begin again with only linearly independent variables, and avoid degeneracy.

The problem remains that in Statistics we are handling data, and data are uncertain. Not only do they contain sampling error, but having sampled our data we have to round them (to the number of decimal places or significant figures we – or the default option of our computer package – choose to work to). We may well be in the general situation, where things are non-degenerate, and there are no non-trivial linear dependence relations. *Nevertheless*, there may be *approximate* linear dependence relations. If so, then rounding error may lead us close to degeneracy (or even to it): our problem is then *numerically unstable*. This phenomenon is known as *multicollinearity*.

Multiple Regression is inherently prone to problems of this kind. One reason is that the more regressors we have, the more ways there are for some of them to be at least approximately linearly dependent on others. This will then cause the problems mentioned above. Our best defence against multicollinearity is to be alert to the danger, and in particular to watch for possible approximate linear dependence relations between regressors. If we can identify such, we have made two important gains:

(i) we can avoid the numerical instability associated with multicollinearity, and reduce the dimension and thus the computational complexity,

(ii) we have identified important structural information about the problem by identifying an approximate link between regressors.

The problem of multicollinearity in fact bedevils the whole subject of Multiple Regression, and is surprisingly common. It is one reason why the subject is 'an art as well as a science'. It is also a reason why automated computer procedures such as the S-Plus commands step and update produce different outcomes depending on the *order* in which variables are declared in the model.

Example 7.6 (Concrete example)

The following example is due to Woods et al. (1932). It is a very good illustration of multicollinearity and how to handle it.

In a study of the production of concrete, the response variable Y is the amount of heat (calories per gram) released while the concrete sets. There are four regressors $X_1, \ldots, X_4$ representing the percentages (by weight rounded to the nearest integer) of the chemically relevant constituents from which the concrete is made. The data are shown in Table 7.1 below.

n	Y	X_1	X_2	X_3	X_4
1	78.5	7	26	6	60
2	74.3	1	29	15	52
3	104.3	11	56	8	20
4	87.6	11	31	8	47
5	95.9	7	52	6	33
6	109.2	11	55	9	22
7	102.7	3	71	17	6
8	72.5	1	31	22	44
9	93.1	2	54	18	22
10	115.9	21	47	4	26
11	83.8	1	40	23	34
12	113.3	11	66	9	12
13	109.9	10	68	8	12

Table 7.1 Data for concrete example

Here the X_i are not exact percentages, due to rounding error and the presence of between 1% and 5% of other chemically relevant compounds. However, X_1, X_2, X_3, X_4 are rounded percentages and so sum to near 100 (cf. the mixture models of Exercise 6.10). So, strong (negative) correlations are anticipated, and we expect that we will not need all of $X_1, \ldots, X_4$ in our chosen model. In this simple example we can fit models using all possible combinations of variables and the results are shown in Table 7.2. Here we cycle through, using as an intuitive guide the proportion of the variability in the data explained by each model as defined by the R^2 statistic (see Chapter 3).

Model	$100R^2$	Model	$100R^2$	Model	$100R^2$
X_1	53.29	$X_1\,X_2$	97.98	$X_1\,X_2\,X_3$	98.32
X_2	66.85	$X_1\,X_3$	54.68	$X_1\,X_2\,X_4$	98.32
X_3	28.61	$X_1\,X_4$	97.28	$X_1\,X_3\,X_4$	98.2
X_4	67.59	$X_2\,X_3$	84.93	$X_2\,X_3\,X_4$	97.33
		$X_2\,X_4$	68.18	$X_1\,X_2\,X_3\,X_4$	98.32
		$X_3\,X_4$	93.69		

Table 7.2   All-subsets regression for Example 7.6

The multicollinearity is well illustrated by the fact that omitting either X_3 or X_4 from the full model does not seem to have much of an effect. Further, the models with just one term do not appear sufficient. Here the t-tests generated as standard output in many computer software packages, in this case R[®][1] using the `summary.lm` command, prove illuminating. When fitting the full model X_1 X_2 X_3 X_4 we obtain the output in Table 7.3 below:

Coefficient	Estimate	Standard Error	t-value	p-value
Intercept	58.683	68.501	0.857	0.417
X_1	1.584	0.728	2.176	0.061
X_2	0.552	0.708	0.780	0.458
X_3	0.134	0.738	0.182	0.860
X_4	-0.107	0.693	-0.154	0.882

Table 7.3 R output for Example 7.6

So despite the high value of R^2, tests for individual model components in the model are non-significant. This in itself suggests possible multicollinearity. Looking at Table 7.2, model selection appears to come down to a choice between the best two-term model X_1 X_2 and the best three-term models X_1 X_2 X_3 and X_1 X_2 X_4. When testing X_1 X_2 X_3 versus X_1 X_2 we get a t-statistic of 0.209 for X_3 suggesting that X_3 can be safely excluded from the model. A similar analysis for the X_1 X_2 X_4 gives a p-value of 0.211 suggesting that X_4 can also be safely omitted from the model. Thus, X_1 X_2 appears to be the best model and the multicollinearity inherent in the problem suggests that a model half the

---

[1] R[®]: A language and environment for statistical computing. © 2009 R Foundation for Statistical Computing, Vienna, Austria. ISBN 3-900051-07-0 http://www. R-project.org

size of the full model will suffice. In larger problems one might suggest using stepwise regression or backward selection starting with the full model, rather than the all-subsets regression approach we considered here.

Regression Diagnostics. A regression analysis is likely to involve an iterative process in which a range of plausible alternative models are examined and compared, before our final model is chosen. This process of *model checking* involves, in particular, looking at unusual or suspicious data points, deficiencies in model fit, etc. This whole process of model examination and criticism is known as *Regression Diagnostics*. For reasons of space, we must refer for background and detail to one of the specialist monographs on the subject, e.g. Atkinson (1985), Atkinson and Riani (2000).

EXERCISES

7.1. Revisit the concrete example using,
   (i) stepwise selection starting with the full model,
   (ii) backward selection starting with the full model,
   (iii) forward selection from the null constant model.

7.2. *Square root transformation for count data.* Counts of rare events are often thought to be approximately Poisson distributed. The transformation $\sqrt{Y}$ or $\sqrt{Y+1}$, if some counts are small, is often thought to be effective in modelling count data. The data in Table 7.4 give a count of the number of poppy plants in oats.
   (i) Fit an Analysis of Variance model using the raw data. Does a plot of residuals against fitted values suggest a transformation?
   (ii) Interpret the model in (i).
   (iii) Re-fit the model in (i-ii) using a square–root transformation. How do your findings change?

Treatment	A	B	C	D	E
Block 1	438	538	77	17	18
Block 2	442	422	61	31	26
Block 3	319	377	157	87	77
Block 4	380	315	52	16	20

Table 7.4 Data for Exercise 7.2

7.3. *Arc sine transformation for proportions.* If we denote the empirical proportions by $\hat{p}$, we replace $\hat{p}$ by introducing the transformation $y = \sin^{-1}(\sqrt{\hat{p}})$. In this angular scale proportions near zero or one are spread out to increase their variance and make the assumption of homogenous errors more realistic. (With small values of $n < 50$ the suggestion is to replace zero or one by $\frac{1}{4n}$ or $1 - \frac{1}{4n}$.) The data in Table 7.5 give the percentage of unusable ears of corn.
(i) Fit an Analysis of Variance model using the raw data. Does a plot of residuals against fitted values suggest a transformation?
(ii) Interpret the model in (i).
(iii) Re-fit the model in (i–ii) using the suggested transformation. How do your findings change?

Block	1	2	3	4	5	6
Treatment A	42.4	34.4	24.1	39.5	55.5	49.1
Treatment B	33.3	33.3	5.0	26.3	30.2	28.6
Treatment C	8.5	21.9	6.2	16.0	13.5	15.4
Treatment D	16.6	19.3	16.6	2.1	11.1	11.1

Table 7.5 Data for Exercise 7.3

7.4. The data in Table 7.6 give the numbers of four kinds of plankton caught in different hauls.
(i) Fit an Analysis of Variance model using the raw data. Does a plot of residuals against fitted values suggest a transformation of the response?
(ii) Calculate the mean and range $(\max(y) - \min(y))$ for each species and repeat using the logged response. Comment.
(iii) Fit an Analysis of Variance model using both raw and logged numbers, and interpret the results.

7.5. Repeat Exercise 7.4 using
(i) The square-root transformation of Exercise 7.2.
(ii) Taylor's power law.

7.6. *The delta method: Approximation formulae for moments of transformed random variables.* Suppose the random vector U satisfies $E(U) = \mu$, $\text{var}(U) = \Sigma_U$, $V = f(U)$ for some smooth function f. Let F_{ij} be the matrix of derivatives defined by

$$F_{ij}(u) = \left(\frac{\partial u}{\partial v}\right)_{ij} = \left(\frac{\partial f}{\partial v}\right)_{ij} = \frac{\partial f_i}{\partial v_j}.$$

Haul	Type I	Type II	Type III	Type IV
1	895	1520	43300	11000
2	540	1610	32800	8600
3	1020	1900	28800	8260
4	470	1350	34600	9830
5	428	980	27800	7600
6	620	1710	32800	9650
7	760	1930	28100	8900
8	537	1960	18900	6060
9	845	1840	31400	10200
10	1050	2410	39500	15500
11	387	1520	29000	9250
12	497	1685	22300	7900

Table 7.6 Data for Exercise 7.4

We wish to construct simple estimates for the mean and variance of V. Set

$$V \approx f(\mu) + F(\mu)(u - \mu).$$

Taking expectations then gives

$$E(V) \approx f(\mu).$$

(i) Show that $\Sigma_V \approx F(\mu)\Sigma_U F(\mu)^T$.

(ii) Let $U \sim Po(\mu)$ and $V = \sqrt{U}$. Give approximate expressions for the mean and variance of V.

(iii) Repeat (ii) for $V = \log(U + 1)$. What happens if $\mu \gg 1$?

7.7. Show, using the delta method, how you might obtain parameter estimates and estimated standard errors for the power-law model $y = \alpha x^{\beta}$.

7.8. *Analysis using graphics in S-Plus/R®*. Re-examine the plots shown in Figures 7.2 and 7.3. The R®-code which produced these plots is shown below. What is the effect of the commands `xaxt/yaxt="n"`? Use `?par` to see other options. Experiment and produce your own examples to show funnelling out and funnelling in of residuals.

Code for funnels out/in plot

```
y2<-(x2+rnorm(60, 0, 0.7))∧2/y2<-(1+x2+rnorm(60, 0,
     0.35))∧0.5
a.lm<-lm(y2~x2)
plot(y2-a.lm$resid, a.lm$resid, xaxt'"n", yaxt="n",
     ylab="Residual", xlab="Fitted value")
```

7.9. For the simple linear model in Exercise 1.6, calculate leverage, Cook's distances, residuals, externally studentised residuals and internally studentised residuals.

7.10. Revisit the simulated data example in Exercise 3.4 using techniques introduced in this chapter.

8
Generalised Linear Models

8.1 Introduction

In previous chapters, we have studied the model

$$y = A\beta + \epsilon,$$

where the mean $Ey = A\beta$ depends linearly on the parameters β, the errors are normal (Gaussian), and the errors are additive. We have also seen (Chapter 7) that in some situations, a transformation of the problem may help to correct some departure from our standard model assumptions. For example, in §7.3 on variance-stabilising transformations, we transformed our data from y to some function $g(y)$, to make the variance constant (at least approximately). We did not there address the effect on the *error structure* of so doing. Of course, $g(y) = g(A\beta + \epsilon)$ as above will *not* have an additive Gaussian error structure any more, even approximately, in general.

   The function of this chapter is to generalise linear models beyond our earlier framework, so as to broaden our scope and address such questions. The material is too advanced to allow a full treatment here, and we refer for background and detail to the (numerous) references cited below, in particular to McCullagh and Nelder (1989) and to Venables and Ripley (2002), Ch. 7.

   We recall that in earlier chapters the Method of Least Squares and the Method of Maximum Likelihood were equivalent. When we go beyond this framework, this convenient feature is no longer present. We use the Method of Maximum Likelihood (equivalent above to the Method of Least Squares, but no

N.H. Bingham and J.M. Fry, *Regression: Linear Models in Statistics,*
Springer Undergraduate Mathematics Series, DOI 10.1007/978-1-84882-969-5_8,
© Springer-Verlag London Limited 2010

longer so in general). This involves us in finding the maximum of the likelihood L, or equivalently the log-likelihood $\ell := \log L$, by solving the *likelihood equation*

$$\ell' = 0.$$

Unfortunately, this equation will no longer have a solution in closed form. Instead, we must proceed as we do when solving a transcendental (or even algebraic) equation

$$f(x) = 0,$$

and proceed numerically. The standard procedure is to use an *iterative* method: to begin with some starting value, x_0 say, and improve it by finding some better approximation x_1 to the required root. This procedure can be iterated: to go from a current approximation x_n to a better approximation x_{n+1}. The usual method here is *Newton–Raphson iteration* (or the *tangent method*):

$$x_{n+1} := x_n - f(x_n)/f'(x_n).$$

This effectively replaces the graph of the function f near the point $x = x_n$ by its tangent at x_n. In the context of statistics, the derivative ℓ' of the log-likelihood function is called the *score function*, s, and the use of iterative methods to solve the likelihood equation is called *Fisher's method of scoring* (see e.g. Kendall and Stuart (1979), §18.21).

Implementation of such an iterative solution by hand is highly laborious, and the standard cases have been programmed and implemented in statistical packages. One consequence is that (at least at the undergraduate level relevant here) in order to implement procedures involving Generalised Linear Models (GLMs), one really needs a statistical package which includes them. The package GLIM®[1] is designed with just this in mind (Aitkin *et al.* (1989), or Crawley (1993)), and also GenStat®[2] (McConway et al. (1999)). For S-Plus for GLMs, we refer to Venables and Ripley (2002), Ch. 7, Crawley (2002), Ch. 27. Unfortunately, the package Minitab® (admirably simple, and very useful for much of the material of this book) does not include GLMs.

Generalised Linear Models, or GLMs, arise principally from the work of the English statistician John A. Nelder (1924–2010); the term is due to Nelder and Wedderburn in 1972; the standard work on the subject is McCullagh and Nelder (1989). As noted above, GLMs may be implemented in GLIM® or GenStat®; the relevant command in S-Plus/R® is `glm`, with the family of error distributions specified, as well as the regressors; see below for examples.

---

[1] GLIM® is a registered trademark of The Royal Statistical Society.

[2] GenStat® is a registered trademark of VSN International Limited, 5 The Waterhouse, Waterhouse Street, Hemel Hempstead, HP1 1ES, UK.

8.2 Definitions and examples

Just as with a linear model, we have regressors, or stimulus variables, $x_1, \ldots, x_p$ say, and a response variable y, which depends on these via a *linear predictor*

$$\eta = \beta_1 x_1 + \ldots + \beta_p x_p,$$

where the β_i are parameters. The mean $\mu = Ey$ depends on this linear predictor η, but whereas in the linear case $\mu = \eta$, we now allow μ to be some smooth invertible function of η, and so also, η is a smooth invertible function of μ. We write

$$\mu = m(\eta), \qquad \eta = m^{-1}(\mu) = g(\mu),$$

where the function g is called the *link function* – it links the linear predictor to the mean. In the linear case, the link g is the identity; we shall see a range of other standard links below.

To complete the specification of the model, we need the distribution of the response variable y, not just its mean μ; that is, we need to specify the *error structure*. We assume that each observation y_i is *independent* and has a density f of the form

$$\exp\left\{\frac{\omega_i(y_i\theta_i - b(\theta_i))}{\phi} + c(y, \phi)\right\},$$

where the parameter θ_i depends on the linear predictor η, ϕ is a scale parameter (which may or may not be known), the ω_i are a sequence of known weights, and $b(.)$ and $c(.)$ are functions. It is further assumed that

$$\mathrm{var}(y_i) = \frac{\phi}{\omega_i} V(\mu_i),$$

where $V(\cdot)$ is a variance function relating the variance of the y_i to the mean μ_i. It can be shown that in the notation above

$$
\begin{aligned}
E(y_i) &= b'(\theta_i), \\
\mathrm{var}(y_i) &= \frac{\phi}{\omega_i} b''(\theta_i).
\end{aligned}
$$

This functional form derives from the theory of *exponential families*, which lies beyond the scope of this book. For a monograph treatment, see e.g. Brown (1986). Suffice it here to say that the parametric families which have a fully satisfactory inference theory are the exponential families. So the assumption above is not arbitrary, but is underpinned by this theory, and GLMs are tractable because of it.

The case when

$$\theta = \eta$$

is particularly important. When it occurs, the link function is called *canonical*. (See also Exercise 8.1).

Example 8.1 (Canonical forms)

1. *Normal.* Here $f(y; \theta, \phi)$ is given by

$$\frac{1}{\sqrt{2\pi\sigma^2}} \exp\{-\frac{1}{2}(y-\mu)^2/\sigma^2\} = \exp\{(y\mu - \mu^2/2)/\sigma^2 - \frac{1}{2}(y^2/\sigma^2 + \log(2\pi\sigma^2))\}.$$

So $\theta = \mu$, the scale parameter is simply the variance σ^2, and the link function g is the *identity* function:

$$g(\mu) = \mu.$$

This of course merely embeds the general linear model, with normal error structure, into the *generalised* linear model as a special case, and was to be expected. The normal distribution is the obvious choice – the 'default option' – for measurement data on the whole line.

2. *Poisson.* Here the mean μ is the Poisson parameter λ, and $f(k; \lambda) = e^{-\lambda}\lambda^k/k!$. Writing y for k to conform with the above,

$$f(y; \lambda) = \exp\{y \log \lambda - \lambda - \log y!\}.$$

So $\theta = \log \lambda$. So the canonical link, when $\theta = \eta = \log \lambda$, is the *logarithm*:

$$\eta = \log \lambda.$$

This explains the presence of the logarithm in §8.3 below on *log-linear models*. The Poisson distribution is the default option for count data (on the non-negative integers). Note also that in this case the scale parameter ϕ is simply $\phi = 1$.

3. *Gamma.* The gamma density $\Gamma(\lambda, \alpha)$ is defined, for parameters α, $\lambda > 0$, as

$$f(x) = \frac{\lambda^\alpha}{\Gamma(\alpha)} e^{-\lambda x} x^{\alpha-1}.$$

The mean is

$$\mu = \alpha/\lambda,$$

and as

$$\begin{aligned} f(x) &= \exp\{-\lambda x + (\alpha - 1)\log x + \alpha \log \lambda - \log \Gamma(\alpha)\} \\ &= \exp\left\{(-\alpha)\frac{x}{\mu} + \ldots\right\}, \end{aligned}$$

the canonical link is the inverse function:

$$\eta = 1/\mu,$$

and we can also read off that the scale parameter is given by

$$\phi = 1/\alpha.$$

The gamma density is the default option for measurement data on the positive half-line. It is often used with the log-link

$$\eta = \log \mu$$

and we shall meet such examples below (see Exercises 8.7).

Other standard examples, included in S-Plus/R®, are the inverse Gaussian family (Exercise 8.9), the binomial (whose special case the Bernoulli, for binary data, we discuss below in §8.3), and the logit, probit and complementary log-log cases (see §8.3 also).

One other pleasant feature of the general linear (normal) case that does not carry over to GLMs is the distribution theory – independent sums of squares, chi-square distributed, leading to F-tests and Analysis of Variance. The distribution theory of GLMs is less simple and clear-cut. Instead of Analysis of Variance, one has *analysis of deviance*. This gives one a means of assessing model fit, and of comparing one model with another – and in particular, of choosing between two or more nested models. For further background and detail, we refer to McCullagh and Nelder (1989), Venables and Ripley (2002), Ch. 7, but we outline the basic procedures in the following two subsections.

8.2.1 Statistical testing and model comparisons

The *scaled deviance* metric is a measure of the distance between the observed y_i and the fitted $\hat{\mu}_i$ of a given model, and is defined as

$$
\begin{aligned}
S(y, \hat{\mu}) &= 2 \left(l(y; \phi, y) - l(\hat{\mu}; \phi, y) \right), \\
&= \frac{2}{\phi} \sum_i \omega_i \left[y_i \left(\theta(y_i) - \hat{\theta}_i \right) - \left(b(\theta(y_i)) - b\left(\hat{\theta}_i\right) \right) \right],
\end{aligned}
$$

where l denotes log-likelihood. We define the *residual deviance* or *deviance* which is the scaled deviance multiplied by the scale parameter ϕ:

$$
D(y, \hat{\mu}) = \phi S(y, \hat{\mu}) = 2 \sum_i \omega_i \left[y_i \left(\theta(y_i) - \hat{\theta}_i \right) - \left(b(\theta(y_i)) - b\left(\hat{\theta}_i\right) \right) \right].
$$

Both the scaled deviance and the residual deviance are important and enable both statistical testing of hypotheses and model comparisons. (Note that the scaled deviance retains the scale parameter ϕ, which is then eliminated from the residual deviance by the above.)

Example 8.2

In the case of the normal linear model, the residual deviance leads to the residual sum of squares:

$$D(y, \hat{\mu}) = SSE = \sum_i (y_i - \hat{\mu})^2.$$

To see this we note that, written as a function of the μ_i, the log-likelihood function is

$$l(\mu|\phi, y) = \frac{1}{2\phi} \sum_i (y_i - \mu_i)^2 + C,$$

where C is constant with respect to μ. We have that

$$D(\hat{\mu}|\phi, y) = 2\phi \left[\frac{-\sum(y_i - y_i)^2 + \sum(y_i - \hat{\mu}_i)^2}{2\phi} \right] = \sum(y_i - \hat{\mu}_i)^2.$$

The residual deviance can also be calculated for a range of common probability distributions (see Exercise 8.2).

Nested models. Nested models can be formally compared using generalised likelihood ratio tests. Suppose Model 1 is $\eta = X\beta$ and Model 2 is $\eta = X\beta + Z\gamma$ with rank$(Z) = r$. Model 1 has dimension p_1 and Model 2 has dimension $p_2 = p_1 + r$. The test statistic is

$$
\begin{aligned}
2(l_2 - l_1) &= S(y; \hat{\mu}_1) - S(y; \hat{\mu}_2), \\
&= \frac{D(y; \hat{\mu}_1) - D(y; \hat{\mu}_2)}{\phi}.
\end{aligned}
$$

If the scale parameter ϕ is known, then the asymptotic distribution of this test statistic should be χ_r^2. This likelihood ratio test also suggests an admittedly rough measure of absolute fit by comparing the residual deviance to χ_{n-p}^2, with high values indicating lack of fit. If ϕ is unknown, one suggestion is to estimate ϕ using Model 2 and then treat ϕ as known. Alternatively, it is often customary to use the F-test

$$\frac{D(y; \hat{\mu}_1) - D(y; \hat{\mu}_2)}{\hat{\phi}r} \sim F_{r, n-p_2},$$

by analogy with the theory of Chapter 6. However, this must be used with caution in non-Gaussian cases. A skeleton analysis of deviance is outlined in Table 8.1, and should proceed as follows:

(i) Test $S(y; \mu_2)$ versus χ_{n-p-1}^2 for an admittedly rough test of model accuracy for model 2.

(ii) Test $S(y; \mu_1) - S(y; \mu_2)$ versus χ_r^2 to test the hypothesis $Z = 0$.

Source	Scaled Deviance	df
Model 2 after fitting Model 1	$S(y; \mu_1)$-$S(y; \mu_2)$	r
Model 2	$S(y; \mu_2)$	$n - p_1 - r$
Model 1	$S(y; \mu_1)$	$n - p_1$

Table 8.1 Skeleton analysis of deviance

Usually more than two models would be compared in the same way. The reader should also note that methods of model selection similar to those discussed in Chapter 6 – namely forward and backward selection and sequential methods – also apply here.

t-tests. Approximate *t*-tests for individual parameters can be constructed by comparing

$$T = \frac{\hat{\beta}_j - \beta_j}{\text{e.s.e}(\hat{\beta}_j)}$$

to t_{n-p} where $\hat{\beta}_j$ is the estimate of β_j and e.s.e denotes the associated estimated standard error. This is partly by analogy with the theory of the Gaussian linear model but also as a way of treating a near-Gaussian situation more robustly. Approximate inference can also be conducted using the delta method of Exercise 7.6. Whilst useful in model simplification, tests based on analysis of deviance are usually preferred when testing between different models. *Non-nested* models may be compared using the following generalisation of AIC:

$$\text{AIC}(\hat{\mu}) = D(y; \hat{\mu}) + 2p\hat{\phi},$$

where μ denotes the fitted values and p the number of parameters of a given model.

8.2.2 Analysis of residuals

There are four types of residuals commonly encountered in Generalised Linear Models and roughly analogous to the various types of residuals defined for the general linear model in Chapter 7. The *response* or *raw* residuals are simply given by

$$e_i = y_i - \hat{\mu}_i.$$

The *Pearson* residuals are defined as

$$e_{P,i} = \sqrt{\omega_i} \frac{y_i - \hat{\mu}_i}{\sqrt{V(\hat{\mu}_i)}} = \sqrt{\hat{\phi}} \frac{y_i - \hat{\mu}_i}{\sqrt{\hat{V}(y_i)}},$$

since $\text{var}(y_i) = (\phi/\omega_i)V(\mu_i)$ by assumption. This is simply $(y_i - \hat{\mu}_i)/\sqrt{\hat{V}(y_i)}$ appropriately scaled so as to remove the dispersion parameter ϕ. A Pearson χ^2 statistic can be defined as

$$\chi^2 = \chi^2(y, \hat{\mu}) = \sum e_{P,i}^2,$$

and can be shown to be asymptotically equivalent to the deviance D. *Working residuals* are defined as

$$e_{W,i} = \frac{(y_i - \hat{\mu}_i)}{d\mu_i/d\eta_i},$$

and are derived as part of the iterative model fitting process. *Deviance residuals* are defined as

$$e_{D,i} = \text{sgn}(y_i - \hat{\mu}_i)2\omega_i \left[y_i \left(\theta(y_i) - \hat{\theta}_i \right) - \left(b(\theta(y_i)) - b\left(\hat{\theta}_i \right) \right) \right],$$

where the sign function sgn (or signum) is defined by

$$sgn(x) = \begin{cases} -1 & x < 0 \\ 0 & x = 0 \\ 1 & x > 0. \end{cases}$$

This definition ensures that $\sum e_{D,i}^2 = D$. If ϕ is not equal to one, the residuals may be multiplied by $\sqrt{\phi}$ or its estimate to produce *scaled* versions of these residuals. Plots of residuals can be used in the usual way to check model adequacy – testing for nonlinearity, outliers, autocorrelation, etc – by plotting against individual covariates or against the $\hat{\mu}_i$ or the $\hat{\eta}_i$. However, in contrast to the general linear model, a Normal probability plot of residuals is unlikely to be helpful. Also, aspects of the data, e.g. Poisson data for small counts, may cause naturally occurring patterns in the residuals which should not then be interpreted as indicating model inadequacy.

8.2.3 Athletics times

Example 8.3

We give a further illustrative example of a gamma Generalised Linear Model by returning to our discussion of athletics times. For distance races, speed decreases with distance, and so the time t taken increases faster than the distance d. Because there are no natural units here, of distance or time, and the relationship between t and d is smooth, *Fechner's Law* applies (Gustav Fechner (1801–1887) in 1860), according to which the relationship should be a *power law*:

$$t = ad^b$$

(see e.g. Hand (2004), §5.6, where it is attributed to Stevens). Here a is *pace*, or time per unit distance (traditionally reckoned in minutes and seconds per mile, or per kilometre), and so is an indicator of the quality of the athlete, while b is dimensionless (and is thought to be much the same for all athletes – see Bingham and Rashid (2008) for background). This is an instance of *Buckingham's Pi Theorem* (Edgar Buckingham (1867–1940) in 1914), according to which a physically meaningful relationship between n physical variables, k of which are independent, can be expressed in terms of $p = n - k$ dimensionless quantities; here $n = 3$ (t, a, d), $k = 2$ (t, d), $p = 1$ (b).

Taking this relationship for the mean $t = ET$ for the actual running time T, one has

$$t = ET = ad^b, \qquad \log(ET) = \log a + b \log d = \alpha + b \log d,$$

say, giving a linear predictor (in $(1, \log d)$) with coefficients α, b. This gives the systematic part of the model; as $\eta = \log \mu$ (with $\mu = ET$ the mean), the link function is log. As time and distance are positive, we take the random part of the model (or error law) as Gamma distributed:

$$T \sim \Gamma(\lambda, \mu).$$

An alternative would be to use an ordinary linear model with Gaussian errors, as in Chapter 3:

$$\log T = \alpha + b \log d + \epsilon, \qquad \epsilon \sim N(0, \sigma^2).$$

With age also present, one needs an age-dependent version of the above: using c in place of a above,

$$ET = c(a)t^b,$$

where in view of our earlier studies one uses a linear model for $c(a)$:

$$Ec(a) = \alpha_1 + \alpha_2 a.$$

The resulting compound model is of *hierarchical* type, as in Nelder, Lee and Pawitan (2006). Here, an approximate solution is possible using the simpler gamma Generalised Linear Model if instead we assume

$$\log(ET) = \alpha_1 + \alpha_2 \log a + b \, \log d.$$

In this case we can use a Gamma Generalised Linear Model with log-link. In S-Plus/R® the relevant syntax required is

```
m1.glm<-glm(time~log(age)+log(distance), family=Gamma(link="log"))
summary(m1.glm)
```

The results obtained for the marathon/half-marathon data (Table 1.1, Exercise 1.3) are shown in Table 8.2, and give similar results to those using a log-transformation and a normal linear model in Example 3.37. As there, the log(age) value of about $1/3$ is consistent (for age$\sim$60, $ET\sim$180) with the Rule of Thumb: expect to lose a minute a year on the marathon through ageing alone.

	Value	Std. Error	t value
Intercept	0.542	0.214	2.538
log(age)	0.334	0.051	6.512
log(distance)	1.017	0.015	67.198

Table 8.2 Regression results for Example 8.3

8.3 Binary models

Logits.

Suppose that we are dealing with a situation where the response y is success or failure (or, life or death) or of zero-one, or Boolean, type. Then if

$$Ey = p,$$

$p \in [0, 1]$, and in non-trivial situations, $p \in (0, 1)$. Then the relevant distribution is *Bernoulli*, with *parameter p, $B(p)$*:

$$p = P(y = 1), \qquad q := 1 - p = P(Y = 0),$$

$$\mathrm{var}(y) = pq = p(1 - p).$$

Interpreting p as the probability of success and $q = 1 - p$ as that of failure, the *odds* on success are $p/q = p/(1 - p)$, and the *log-odds*, more natural from some points of view, are

$$\log\left(\frac{p}{1 - p}\right).$$

Thinking of success or failure as survival or death in a medical context of treatment for some disease, the log-odds for survival may depend on covariates: age might well be relevant, so too might length of treatment, how early the disease was diagnosed, treatment type, gender, blood group etc. The simplest plausible model is to assume that the log-odds of survival depend on some *linear predictor η* – a linear combination $\eta = \sum_j a_j \beta_j$ of parameters β_j, just as before (cf. §9.5 below on survival analysis). With data $y_1, \ldots, y_n$ as before, and writing

$$Ey_i = p_i \qquad (i = 1, \ldots, n),$$

we need a double-suffix notation just as before, obtaining

$$\log\{p_i/(1 - p_i)\} = \sum_{j=1}^{p} a_{ij}\beta_j, \qquad (i = 1, \ldots, n).$$

There are three salient features here:
(i) The function

$$g(p) = \log\{p/(1 - p)\},$$

the link function, which links mean response $p = Ey$ to the linear predictor.
(ii) The *distributions* ('error structure'), which belong to the *Bernoulli* family $B(p)$, a special case of the *binomial* family $B(n, p)$, under which

$$P(X = k) = \binom{n}{k} p^k (1 - p)^{n-k} \qquad (k = 0, 1, \ldots, n).$$

(iii) The function V giving the variance in terms of the mean:

$$V(p) = p(1 - p),$$

called the *variance function*.

The model above is called the *logit* model (from log-odds), or *logistic* model (as if $\eta = \log\{p(1 - p)\}$, $p = e^{\eta}/(1 + e^{\eta})$, the logistic function). Binary data are very important, and have been studied at book length; see e.g. McCullagh and Nelder (1989) Ch. 13, Cox and Snell (1989), and Collett (2003). The relevant S-Plus/R® commands are of the form

```
glm(y ~ ..., family = binomial)
```

We draw an illustrative example (as usual) from athletics times. The 'time to beat' for a club runner of reasonable standard in the marathon is three hours; let us interpret 'success' as breaking three hours. The sample version of the expected frequency p of success is the observed frequency, the proportion of successful runners. For a mass event (such as the London Marathon), which we suppose for simplicity has reached a steady state in terms of visibility, prestige etc., the systematic component of the observed variability in frequency of success from year to year is governed principally by the weather conditions: environmental factors such as temperature, humidity, wind and the like. At too high a temperature, the body is prone to dehydration and heat-stroke; at too low a temperature, the muscles cannot operate at peak efficiency. Performance thus suffers on either side of the optimum temperature, and a quadratic in temperature is suggested. On the other hand, humidity is simply bad: the more humid the air is, the harder it is for sweat to evaporate – and so perform its function, of cooling the body (heat is lost through evaporation). In an endurance event in humid air, the body suffers doubly: from fluid loss, and rise in core temperature. Thus a linear term in humidity is suggested.

Probits.

A very different way of producing a mean response in the interval $(0, 1)$ from a linear predictor is to apply the (standard) normal probability distribution function Φ. The model

$$p = \Phi(\alpha + \beta x)$$

(or some more complicated linear predictor) arises in bioassay, and is called a *probit* model. Writing $\eta = \sum_j \beta_j x_j$ for the linear predictor, the link function is now

$$\eta = g(p) = \Phi^{-1}(p).$$

Complementary log-log link.

In dilution assay, the probability p of a tube containing bacteria is related to the number $x = 0, 1, 2, \ldots$ of dilutions by

$$p = 1 - e^{-\lambda x}$$

for some parameter λ (the number of bacteria present is modelled by a Poisson distribution with this parameter). The link function here is

$$\eta = g(p) = \log(-\log(1 - p)) = \log \lambda + \log x.$$

Example 8.4

The data in Table 8.3 show the number of insects killed when exposed to different doses of insecticide.

Dose	Number	Number killed	% killed
10.7	50	44	88
8.2	49	42	86
5.6	46	24	52
4.3	48	16	33
3.1	50	6	12
0.5	49	0	0

Table 8.3 Data for Example 8.4

We wish to model these data using a Generalised Linear Model. A sensible starting point is to plot the empirical logits defined here as $\eta_{e,i} =$

$\log(y_i + 1/2) - \log(1 - y_i + 1/2)$, where the $1/2$ guards against singularities in the likelihood function if $y_i = 0$ or $y_i = 1$. Here, a plot of the $\eta_{e,i}$ against $\log(\text{dose})$ appears roughly linear suggesting a logarithmic term in dose. The model can be fitted in R® as follows. First, the count data needs to be stored as two columns of successes and failures (the command `cbind` is helpful here). The model is fitted with the following commands:

```
a.glm<-glm(data~log(dose), family=binomial)
summary(a.glm)
```
This gives a residual deviance of 1.595 with 4 df The deviance of the null model with only a constant term is 163.745 on 5 df Testing 1.595 against χ_4^2 gives a p-value of 0.810, so no evidence of lack of fit. The log(dose) term is highly significant. The analysis of deviance test gives $163.745 - 1.595 = 162.149$ on 1 df with $p = 0.000$. Probit and complementary log-log models can be fitted in S-Plus/R® using the following syntax (see Exercise 8.4):

```
a.glm<-glm(data~log(dose), family=binomial(link=probit))
a.glm<-glm(data~log(dose), family=binomial(link=cloglog))
```

8.4 Count data, contingency tables and log-linear models

Suppose we have n observations from a population, and we wish to study a characteristic which occurs in r possible types. We classify our observations, and count the numbers $n_1, \ldots, n_r$ of each type (so $n_1 + \ldots + n_r = n$). We may wish to test the hypothesis H_0 that type k occurs with probability p_k, where

$$\sum_{k=1}^{r} p_k = 1.$$

Under this hypothesis, the expected number of type k is $e_k = np_k$; the observed number is $o_k = n_k$. Pearson's *chi-square goodness-of-fit test* (Karl Pearson (1857–1936), in 1900) uses the *chi-square statistic*

$$X^2 := \sum_{k=1}^{r} (n_k - np_k)^2/(np_k) = \sum (o_k - e_k)^2/e_k.$$

Then for large samples, X^2 has approximately the distribution $\chi^2(r-1)$, the chi-square distribution with r df; large values of X^2 are evidence against H_0. The proof proceeds by using the multidimensional Central Limit Theorem to show that the random vector $(x_1, \ldots, x_r)$, where

$$x_k := (n_k - np_k)/\sqrt{np_k},$$

is asymptotically multivariate normal, with mean zero and (symmetric) covariance matrix

$$A = I - pp^T,$$

where p is the column vector

$$(\sqrt{p_1}, \ldots, \sqrt{p_r})^T.$$

Since $\sum_k p_k = 1$, A is idempotent; its trace, and so its rank, is $r - 1$. This loss of one degree of freedom corresponds to the one linear constraint satisfied (the n_k sum to n; the p_k sum to 1). From this, the limiting distribution $\chi^2(r - 1)$ follows by Theorem 3.16. For details, see e.g. Cramér (1946), §30.1.

Now the distribution of the vector of observations $(n_1, \ldots, n_r)$ (for which $\sum_i n_i = n$) is *multinomial*:

$$P(n_1 = k_1, \ldots, n_r = k_r) = \binom{n}{k_1, \ldots, k_r} p_1^{k_1} \ldots p_r^{k_r},$$

for any non-negative integers $k_1, \ldots, k_r$ with sum n (the multinomial coefficient counts the number of ways in which the k_1 observations of type 1, etc., can be chosen; then $p_1^{k_1} \ldots p_r^{k_r}$ is the probability of observing these types for each such choice.

According to the *conditioning property* of the Poisson process (see e.g. Grimmett and Stirzaker (2001), §6.12–6.13), we obtain multinomial distributions when we condition a Poisson process on the number of points (in some region).

These theoretical considerations lie behind the use of GLMs with *Poisson errors* for the analysis of count data. The basic observation here is due to Nelder in 1974. In the linear model of previous chapters we had additive normal errors, and – regarded as a GLM – the identity link. We now have *multiplicative Poisson errors*, the multiplicativity corresponding to the logarithmic link.

We assume that the logarithm of the ith data point, $\mu_i = Ey_i$, is given by a linear combination of covariates:

$$\log \mu_i = \eta_i = \beta^T \mathbf{x}_i \qquad (i = 1, \ldots, n).$$

We shall refer to such models as *log-linear models*. For them, the link function is the logarithm:

$$g(\mu) = \log \mu.$$

Example 8.5 (Poisson modelling of sequences of small counts)

Suppose that we have the following (artificial) data in Table 8.4 and we wish to model this count data using a Poisson Generalised Linear Model.

x	1	2	3	4	5	6	7	8	9	10	11	12	13	14
y	1	0	2	5	6	9	12	12	25	25	22	30	52	54

Table 8.4 Data for Example 8.5

A plot of the guarded logs, $\log(y_i + 0.5)$, against x_i seems close to a straight line although there is perhaps a slight suggestion of curvature. The model with x on its own gives a residual deviance of 24.672 on 12 df The χ^2 goodness-of-fit test gives a p-value of 0.016, suggesting that the fit of this model is poor. The model with a quadratic term has a residual deviance of 13.986 on 11 df This model seems to fit better; the χ^2 goodness of fit test gives a p-value of 0.234, and the AIC of this model is 75.934. A plot of the guarded logs against $\log(x_i)$ also appears close to linear and $\log(x)$ thus seems a suitable candidate model. Fitting this model gives a residual deviance of 14.526 on 12 df and appears reasonable (χ^2 test gives $p = 0.268$). The AIC for this model is 74.474 and thus $\log(x)$ appears to be the best model.

All of this continues to apply when our counts are cross-classified by more than one characteristic. We consider first the case of *two* characteristics, partly because it is the simplest case, partly because we may conveniently display count data classified by two characteristics in the form of a *contingency table*. We may then, for example, test the null hypothesis that the two characteristics are independent by forming an appropriate chi-square statistic. For large samples, this will (under the null hypothesis) have approximately a chi-square distribution with df $(r-1)(s-1)$, where r and s are the numbers of forms of the two characteristics. For proof, and examples, see e.g. Cramér (1946), Ch. 30.

We may very well have more than two characteristics. Similar remarks apply, but the analysis is more complicated. Such situations are common in the social sciences – sociology, for example. Special software has been developed: SPSS®[3] (statistical package for the social sciences). Such multivariate count data is so important that it has been treated at book length; see e.g. Bishop et al. (1995), Plackett (1974), Fienberg (1980).

Another application area is insurance. A motor insurer might consider, when assessing the risk on a policy, the driver's age, annual mileage, sex, etc; also the type of vehicle (sports cars are often charged higher premiums), whether used for work, whether kept off-road, etc. A house insurer might consider number of rooms (or bedrooms), indicators of population density, postal code (information about soil conditions, and so subsidence risk, for buildings; about the ambient population, and so risk of burglary, for contents, etc.). The simplest

---

[3] SPSS® is a registered trademark of SPSS Inc., 233 S. Wacker Drive, 11th Floor, Chicago, IL 60606, USA, http://www.spss.com

way to use such information is to use a linear regression function, or linear predictor, as above, whence the relevance of GLMs. The S-Plus commands are much as before:

glm(y ~ ..., family = poisson).

We note in passing that the parameter λ in the Poisson distribution $P(\lambda)$, giving its mean and also its variance, is most naturally viewed as a *rate*, or *intensity*, of a stochastic process – the *Poisson point process* with rate λ (in time, or in space) – which corresponds to a *risk* in the insurance context. Thus this material is best studied in tandem with a study of stochastic processes, for which we refer to, e.g., Haigh (2002), Ch. 8, as well as Grimmett and Stirzaker (2001), Ch. 6 cited earlier.

Example 8.6 (Skeleton analysis of 2×2 contingency tables)

For technical reasons, it can be important to distinguish between two cases of interest.

Two response variables. Both variables are random, only the total sample size $\sum_{ij} y_{ij}$ is fixed. The data in Exercise 7.4 are an example with two response variables.

One response variable and one observed variable. The setting here is a controlled experiment rather than an observational study. The design of the experiment fixes row or column totals *before* the full results of the experiment are known. One example of this is medical trials where patients are assigned different treatment groups, e.g. placebo/vaccine, etc. The interested reader is referred to Dobson and Barnett (2003), Ch. 9.

A range of different possible hypotheses applies in each of these two cases. Apart from unrealistic or very uncommon examples, the main interest lies in testing the hypothesis of no association between the two characteristics A and B. It can be shown that this reduces to testing the adequacy of the log-linear model

$$\log(Y) = \text{const.} + A + B.$$

The data in Table 8.5 give hair and eye colours for a group of subjects. We use Poisson log-linear models to test for an association between hair and eye colour. Fitting the model we obtain a residual deviance of 146.44 on 9 df leading to a p-value of 0.000 and we reject the null hypothesis of no association.

	Brown	Blue	Hazel	Green
Black hair	68	20	15	5
Brown hair	119	84	54	29
Red hair	26	17	14	14
Blond hair	7	94	10	16

Table 8.5 Data for Example 8.6

8.5 Over-dispersion and the Negative Binomial Distribution

The fact that a Poisson mean and variance coincide gives a yardstick by which to judge variability, or dispersion, of count data. If the variance-to-mean ratio observed is > 1, the data are called *over-dispersed* (if < 1, they are called *under-dispersed*, though this is less common). Equivalently, one may also use the ratio of standard error to mean (coefficient of variation), often preferred to the variance-mean ratio as it is dimensionless.

One model used for over-dispersion is to take a *Gamma mixture of Poissons*: take a Poisson distribution with random mean, M say, where M is Gamma distributed. Thus

$$P(Y = n | M = \lambda) = e^{-\lambda} \lambda^n / n!,$$

but (it is convenient here to reparametrise, from λ, $\alpha > 0$ to ν, $\tau > 0$) $M \sim \Gamma(\nu/\tau, \nu)$: M has density

$$f(y) = \frac{1}{\Gamma(\nu)} \left(\frac{\nu y}{\tau}\right)^\nu e^{-\nu y / \tau} \frac{1}{y} \qquad (y > 0).$$

Then unconditionally

$$
\begin{aligned}
P(Y = n) &= \int_0^\infty \frac{e^{-y} y^n}{n!} \frac{1}{\Gamma(\nu)} \left(\frac{\nu y}{\tau}\right)^\nu e^{-\nu y / \tau} y^{n+\nu-1} \, dy \\
&= \frac{\nu^\nu}{\tau^\nu} \frac{1}{n! \Gamma(\nu)} \frac{1}{(1 + \nu/\tau)^{n+\nu}} \int_0^\infty e^{-u} u^{n+\nu-1} \, du \quad (y(1 + \nu/\tau) = u) \\
&= \frac{\nu^\nu}{\tau^\nu (1 + \nu/\tau)^{n+\nu}} \frac{\Gamma(n+\nu)}{n! \Gamma(\nu)}.
\end{aligned}
$$

This is the *Negative Binomial* distribution, $NB(\nu, \tau)$, in one of several parametrisations (compare McCullagh and Nelder (1989), p237 and p373). The mean is

$$\mu = \tau.$$

The variance is

$$V(\mu) = \tau + \tau^2 / \nu = \mu + \mu^2 / \nu.$$

The model is thus over-dispersed.

Since $\Gamma(1 + x) = x\Gamma(x)$,

$$\frac{\Gamma(n + \nu)}{n!\Gamma(\nu)} = \frac{(n + \nu - 1)(n + \nu - 2)\ldots\nu}{n!},$$

and when ν is a positive integer, r say, this has the form of a binomial coefficient

$$\binom{n + r - 1}{n} = \binom{n + r - 1}{r - 1}.$$

In this case,

$$P(Y = n) = \binom{n + r - 1}{n} p^r q^n \qquad (n = 0, 1, \ldots),$$

writing

$$p := r/(\tau + r), \qquad q := 1 - p = \tau/(\tau + r).$$

The case $r = 1$ gives the *geometric* distribution, $G(p)$:

$$P(Y = n) = q^n p \qquad (n = 0, 1, \ldots),$$

the distribution of the number of failures before the first success in Bernoulli trials with parameter p ('tossing a p-coin'). This has mean q/p and variance q/p^2 (over-dispersed, since $p \in (0, 1)$, so $1/p > 1$). The number of failures before the rth success has the negative binomial distribution in the form just obtained (the binomial coefficient counts the number of ways of distributing the n failures over the first $n + r - 1$ trials; for each such way, these n failures and $r - 1$ successes happen with probability $q^n p^{r-1}$; the $(n + r)$th trial is a success with probability p). So the number of failures before the rth success
(i) has the negative binomial distribution (which it is customary and convenient to parametrise as $NB(r, p)$ in this case);
(ii) is the sum of r independent copies of geometric random variables with distribution $G(p)$;
(iii) so has mean rq/p and variance rq/p^2 (agreeing with the above with $r = \nu$, $p = r/(\tau + r)$, $q = \tau/(\tau + r)$).
The Federalist.

The Federalist Papers were a series of essays on constitutional matters, published in 1787–1788 by Alexander Hamilton, John Jay and James Madison to persuade the citizens of New York State to ratify the U.S. Constitution. Authorship of a number of these papers, published anonymously, was later disputed between Hamilton and Madison. Their authorship has since been settled by a classic statistical study, based on the use of the negative binomial distribution for over-dispersed count data (for usage of key indicator words – 'whilst' and 'while' proved decisive); see Mosteller and Wallace (1984).

8.5.1 Practical applications: Analysis of over-dispersed models in R®

For binomial and Poisson families, the theory of Generalised Linear Models specifies that the dispersion parameter $\phi = 1$. Over-dispersion can be very common in practical applications and is typically characterised by the residual deviance differing significantly from its asymptotic expected value given by the residual degrees of freedom (Venables and Ripley (2002)). Note, however, that this theory is only asymptotic. We may crudely interpret over-dispersion as saying that data varies more than if the underlying model really were from a Poisson or binomial sample. A solution is to multiply the variance functions by a dispersion parameter ϕ, which then has to be estimated rather than simply assumed to be fixed at 1. Here, we skip technical details except to say that this is possible using a *quasi-likelihood* approach and can be easily implemented in R® using the Generalised Linear Model families `quasipoisson` and `quasibinomial`. We illustrate the procedure with an application to over-dispersed Poisson data.

Example 8.7

We wish to fit an appropriate Generalised Linear Model to the count data of Exercise 7.2. Fitting the model with both blocks and treatments gives a residual deviance of 242.46 on 12 df giving a clear indication of over-dispersion. A quasi-poisson model can be fitted with the following commands:

```
m1.glm<-glm(data~blocks+treatments, family=quasipoisson)
summary(m1.glm)
```

Since we have to estimate the dispersion parameter ϕ we use an F-test to distinguish between the models with blocks and treatments and the model with blocks only. We have that

$$F = \frac{\Delta\text{Residual deviance}}{\Delta\text{df}(\hat{\phi})} = \frac{3468.5 - 242.46}{4(21.939)} = 36.762.$$

Testing against $F_{4,12}$ gives a p-value of 0.000. Similar procedures can be used to test the effectiveness of blocking (see Exercise 8.5).

EXERCISES

8.1. *Canonical forms.* Show that these common probability distributions can be written in the canonical form of a Generalised Linear Model as shown in Table 8.6:

	Normal $N(\theta, \phi)$	Poisson $\text{Po}(e^\theta)$	Binomial $ny \sim \text{Bi}\left(n, \frac{e^\theta}{1+e^\theta}\right)$	Gamma $\Gamma\left(\frac{1}{\phi}, -\frac{\theta}{\phi}\right)$
$\frac{\phi}{\omega}$	ϕ	1	n^{-1}	ϕ
$b(\theta)$	$\frac{\theta^2}{2}$	e^θ	$\log\left(1+e^\theta\right)$	$-\log(-\theta)$
$c(y, \theta)$	$-\frac{y^2}{2\phi} - \frac{\phi \log(2\pi)}{2}$	$-\log(y!)$	$\log\left(\begin{array}{c} n \\ ny \end{array}\right)$	$\left(\frac{1}{\phi} - 1\right)\log y$ $-\frac{\log \phi}{\phi} + \log\ \phi$
$\mu = b'(\theta)$	θ	e^θ	$\frac{e^\theta}{1+e^\theta}$	$-\frac{1}{\theta}$
$b''(\theta)$	1	μ	$\mu(1-\mu)$	μ^2

Table 8.6 Canonical forms for Exercise 8.1

8.2. *(Residual) deviance calculations.* Show that for the following common probability distributions the residual deviances can be calculated as follows:

Poisson
$$2\sum_i \left(y_i \log\left(\frac{y_i}{\hat{\mu}_i}\right) - (y_i - \hat{\mu}_i)\right),$$

Binomial
$$2\sum_i n_i \left\{y_i \log\left(\frac{y_i}{\hat{\mu}_i}\right) + (1 - y_i)\log\left(\frac{1 - y_i}{1 - \hat{\mu}_i}\right)\right\},$$

Gamma
$$2\sum_i \log\left(\frac{\hat{\mu}_i}{y_i}\right) + \frac{y_i - \hat{\mu}_i}{\hat{\mu}_i}.$$

8.3. Test the hypothesis of no association between haul and number for the data in Exercise 7.4 using
(i) a Poisson log-linear model,
(ii) the Pearson χ^2 test of no association,
and comment on your findings.

8.4. Re-fit the data in Example 8.4 using
(i) a probit model,
(ii) a complementary log-log model,
(iii) an approximate method using general linear models.

8.5. Re-fit the data in Exercise 7.2 using a Poisson Generalised Linear Model, before switching to an over-dispersed Poisson model if this seems appropriate. Test for the effectiveness of blocking by seeing if the model with just the blocks term offers an improvement over the null model.

8.6. Suppose that we have the following data for the number of unusable ears of corn shown in Table 8.7. (Assume totals are out of 36.) Analyse these data by fitting a binomial Generalised Linear Model, using a quasi-binomial model if it appears that we have over-dispersion. Compare your results with an approximation using General Linear Models on similar data in Exercise 7.3 and interpret the results.

Block	1	2	3	4	5	6
Treatment A	15	12	9	14	20	18
Treatment B	12	12	2	9	11	10
Treatment C	3	8	2	6	5	6
Treatment D	6	7	6	1	4	4

Table 8.7 Data for Exercise 8.6

8.7. *Generalised Linear Model with Gamma errors.* Using the data in Exercise 1.6 fit a Gamma Generalised Linear Model. Interpret your findings and compare both with Exercise 1.6 and the analyses in §5.3. Write down the equation of your fitted model.

8.8. *Inverse Gaussian distribution.* The inverse Gaussian distribution is the distribution on the positive half-axis with probability density

$$f(y) = \sqrt{\frac{\lambda}{2\pi y^3}} \exp\left(\frac{-\lambda(y-\mu)^2}{2\mu^2 y}\right).$$

Show that this density lies in the exponential family (see Exercise 8.1).

8.9. *Generalised Linear Model with inverse Gaussian errors.* Repeat Exercise 8.7 using an inverse Gaussian Generalised Linear Model.

8.10. *The effect of ageing on athletic performance.* Using the fitted equations obtained in Exercises 8.7 and 8.9 and using $x = 63$, comment on the effect of
(i) ageing,
(ii) club status.

9
Other topics

9.1 Mixed models

In §5.1 we considered extending our initial model (M_0), with p parameters, to an augmented model M_A with a further q parameters. Here, as in Chapter 2, we have $p + q << n$, there are many fewer parameters than data points. We now turn to a situation with some similarities but with important contrasts. Here our initial model has *fixed effects*, but our augmented model adds *random effects*, which may be comparable in number to the sample size n.

We mention some representative situations in which such mixed models occur.

1. Longitudinal studies (or *panel data*). Suppose we wish to monitor the effect of some educational initiative. One may choose some representative sample or cohort of school children or students, and track their progress over time. Typically, the resulting data set consists of a large number (the size of the cohort) of short time series (the longer the time the more informative the study, but the more expensive it is, and the longer the delay before any useful policy decisions can be made). For background on longitudinal data, see Diggle et al. (2002).

Here one takes for granted that the children in the cohort differ – in ability, and in every other aspect of their individuality. One needs information on between-children variation (that is, on cohort variance); this becomes a parameter in the mixed model. The child effects are the random effects: if one repeated the study with a different cohort, these would be different. The educational aspects one wishes to study are the fixed effects.

N.H. Bingham and J.M. Fry, *Regression: Linear Models in Statistics*,
Springer Undergraduate Mathematics Series, DOI 10.1007/978-1-84882-969-5_9,
© Springer-Verlag London Limited 2010

2. Livestock studies. One may wish to follow the effect of some treatments – a diet, or dietary supplements, say – over time, on a cohort of livestock (cattle, sheep or pigs, say). Again, individual animals differ, and these give the random effects. The fixed effects are the objects of study.

The field of mixed models was pioneered in the US dairy industry by C. R. Henderson (1911–1989) from 1950 on, together with his student S. R. Searle (1928–). Searle is the author of standard works on linear models (Searle (1991)), variance components (Searle, Casella and McCulloch (1992)), and matrix theory for statisticians (Searle (1982)). Henderson was particularly interested in selection of sires (breeding bulls) in the dairy industry. His work is credited with having produced great gains in yields, of great economic value.

3. Athletics times. One may wish to study the effect of ageing on athletes past their peak. One way to do this is to extract from the race results of a particular race over successive years the performances of athletes competing repeatedly. Again, individual athletes differ; these are the random effects. Fixed effects one might be interested in include age, sex and club status. For background, see Bingham and Rashid (2008).

We shall follow the notation of §5.1 fairly closely. Thus we write

$$W = (X, Z)$$

for the new design matrix $(n \times (p + q))$. It is convenient to take the random effects – which as is customary we denote by u – to have zero mean (any additive terms coming from the mean Eu can be absorbed into the fixed effects). Thus the *linear mixed model* is defined by

$$y = X\beta + Zu + \epsilon, \qquad (LMM)$$

where (both means are zero and) the covariance matrices are given by

$$E\epsilon = Eu = 0, \ \text{cov}(\epsilon, u) = 0, \quad R := \text{var } \epsilon, \ D := \text{var } u,$$

('R for regresssion, D for dispersion'). One can write (LMM) as an ordinary linear model,

$$y = X\beta + \epsilon^*, \qquad \epsilon^* := Zu + \epsilon.$$

By Proposition 4.5, this has covariance matrix

$$V := \text{cov } \epsilon^* = ZDZ^T + R$$

('V for variance'). So by Theorem 3.5, the generalised least-squares solution is

$$\hat{\beta} = (X^T V^{-1} X)^{-1} X^T V^{-1} y. \qquad (GLS)$$

We now specify the distributions in our model by assuming that u is multivariate normal (multinormal), and that the conditional distribution of y given u is also multinormal:

$$y|u \sim N(X\beta + Zu, R), \qquad u \sim N(0, D). \qquad (NMM)$$

Then the (unconditional) distribution of y is a *normal mean mixture*, whence the name (NMM). Now the joint density $f(y, u)$ is

$$f(y, u) = f(y|u)f(u),$$

the product of the conditional density of y given u and the density of u. So

$$f(y, u) = const. \exp\{-\frac{1}{2}(y - X\beta - Zu)^T R^{-1}(y - X\beta - Zu)\}. \exp\left\{-\frac{1}{2}u^T D^{-1}u\right\}.$$

Thus to maximise the likelihood (with respect to β and u), we maximise $f(y, u)$, that is, we minimise:

$$\min \qquad (y - X\beta - Zu)^T R^{-1}(y - X\beta - Zu) + u^T D^{-1}u. \qquad (pen)$$

Note the different roles of the two terms. The first, which contains the data, comes from the likelihoood; the second comes from the random effects. It serves as a penalty term (the penalty we pay for not knowing the random effects). So we have here a *penalised likelihood* (recall we encountered penalised likelihood in §5.2.1, in connection with nested models and AIC).

The least-squares solution of Chapters 3, 4 gives the best linear unbiased estimator or BLUE (see §3.3). It is conventional to speak of *predictors*, rather than estimators, with random effects. The solution is thus a *best linear unbiased predictor*, or BLUP.

Theorem 9.1

The BLUPs – the solutions $\hat{\beta}$, $\hat{u}$, of the minimisation problem (MME) – satisfy

$$\left.\begin{array}{rcl} XR^{-1}X\hat{\beta} + X^T R^{-1}Z\hat{u} &=& X^T R^{-1}y, \\ ZR^{-1}X\hat{\beta} + \left[Z^T R^{-1}Z + D^{-1}\right]\hat{u} &=& Z^T R^{-1}y \end{array}\right\} \qquad (MME)$$

(Henderson's mixed model equations of 1950).

Proof

We use the vector calculus results of Exercises 3.6–3.7. If we expand the first term in (pen) above, we obtain nine terms, but the quadratic form in y does not involve β or u, so we discard it; this with the second term above gives nine terms, all scalars, so all their own transposes. This allows us to combine three pairs of terms, reducing to six terms, two linear in β, two linear in u and two cross terms in β and u; there is also a quadratic term in β, and two quadratic terms in u, which we can combine. Setting the partial derivatives with respect to β and u equal to zero then gives

$$-2y^T R^{-1} X + 2u^T Z^T R^{-1} X + 2\beta^T X^T R^{-1} X = 0,$$
$$-2y^T R^{-1} Z + 2\beta^T X^T R^{-1} Z + 2u^T \left[Z^T R^{-1} Z + D^{-1} \right] = 0,$$

or

$$\left. \begin{aligned} X^T R^{-1} X \beta + X^T R^{-1} Z u &= X^T R^{-1} y, \\ Z^T R^{-1} X \beta + [Z^T R^{-1} Z + D^{-1}] u &= Z^T R^{-1} y, \end{aligned} \right\} \qquad (MME)$$

as required. $\qquad \square$

9.1.1 Mixed models and Generalised Least Squares

To proceed, we need some matrix algebra. The next result is known as the *Sherman–Morrison–Woodbury formula*, or *Woodbury's formula* (of 1950).

Lemma 9.2 (Woodbury's Formula)

$$(A + UBV)^{-1} = A^{-1} - A^{-1}U.(I + BVA^{-1}U)^{-1}.BVA^{-1},$$

if all the matrix products are conformable and all the matrix inverses exist.

Proof

We have to show that if we pre-multiply or post-multiply the right by $A + UBV$ we get the identity I.

Pre-multiplying, we get four terms. Taking the first two as those from $(A + UBV)A^{-1}$, these are

$$I + UBVA^{-1} - U(I + BVA^{-1}U)^{-1}BVA^{-1}$$
$$-UBVA^{-1}U(I + BVA^{-1}U)^{-1}BVA^{-1}.$$

The third and fourth terms combine, to give

$$I + UBVA^{-1} - U.BVA^{-1} = I,$$

as required. The proof for post-multiplying is similar. $\qquad \square$

Applied in the context of §9.1 (where now $V := ZDZ^T + R$, as above), this gives

Corollary 9.3

(i)

$$V^{-1} := (ZDZ^T + R)^{-1} = R^{-1} - R^{-1}Z(Z^T R^{-1}Z + D^{-1})^{-1}ZR^{-1}.$$

(ii)

$$DZ^T V^{-1} = (Z^T R^{-1}Z + D^{-1})^{-1}Z^T R^{-1}.$$

Proof

For (i), we use Woodbury's Formula with R, Z, D, Z^T for A, U, B, V:

$$
\begin{aligned}
(R + ZDZ^T)^{-1} &= R^{-1} - R^{-1}Z.(I + DZ^T R^{-1}Z)^{-1}.DZ^T R^{-1} \\
&= R^{-1} - R^{-1}Z.[D(D^{-1} + Z^T R^{-1}Z)]^{-1}.DZ^T R^{-1} \\
&= R^{-1} - R^{-1}Z.(D^{-1} + Z^T R^{-1}Z)^{-1}.Z^T R^{-1}.
\end{aligned}
$$

For (ii), use Woodbury's Formula with D^{-1}, Z^T, R^{-1}, Z for A, U, B, V:

$$(D^{-1} + Z^T R^{-1}Z)^{-1} = D - DZ^T.(I + R^{-1}ZDZ^T)^{-1}.R^{-1}ZD,$$

so

$$(D^{-1} + Z^T R^{-1}Z)^{-1}Z^T R^{-1} = DZ^T R^{-1} - DZ^T(I + R^{-1}ZDZ^T)^{-1}R^{-1}ZDZ^T R^{-1}.$$

The right is equal to $DZ^T[I - (I + R^{-1}ZDZ^T)^{-1}R^{-1}ZDZ^T]R^{-1}$, or equivalently, to $DZ^T[I - (I + R^{-1}ZDZ^T)^{-1}\{(I + R^{-1}ZDZ^T) - I\}]R^{-1}$. Combining, we see that

$$
\begin{aligned}
(D^{-1} + Z^T R^{-1}Z)^{-1}Z^T R^{-1} &= DZ^T[I - I + (I + R^{-1}ZDZ^T)^{-1}]R^{-1} \\
&= DZ^T(R + ZDZ^T)^{-1} \\
&= DZ^T V^{-1},
\end{aligned}
$$

as required.   □

Theorem 9.4

The BLUP $\hat{\beta}$ in Theorem 9.1 is the same as the generalised least-squares estimator:

$$\hat{\beta} = (X^T V^{-1}X)^{-1} X^T V^{-1}y. \qquad (GLS)$$

The BLUP $\hat{u}$ is given by either of

$$\hat{u} = \left(Z^T R^{-1} Z + D^{-1}\right)^{-1} Z^T R^{-1} \left(y - X\hat{\beta}\right)$$

or

$$\hat{u} = D Z^T V^{-1} \left(y - X\hat{\beta}\right).$$

Proof

We eliminate $\hat{u}$ between the two equations (MME). To do this, pre-multiply the second by $X^T R^{-1} Z (Z^T R^{-1} Z + D^{-1})^{-1}$ and subtract. We obtain that

$$X^T R^{-1} X\hat{\beta} - X^T R^{-1} Z \left(Z^T R^{-1} Z + D^{-1}\right)^{-1} Z^T R^{-1} X\hat{\beta} \quad =$$

$$X^T R^{-1} y - X^T R^{-1} Z \left(Z^T R^{-1} Z + D^{-1}\right)^{-1} Z^T R^{-1} y. \qquad (a)$$

Substitute the matrix product on the right of Corollary 9.3(i) into both sides of (a):

$$X^T R^{-1} X\hat{\beta} - X^T \left\{R^{-1} - V^{-1}\right\} X\hat{\beta} = X^T R^{-1} y - X^T \left\{R^{-1} - V^{-1}\right\} y,$$

or

$$X^T V^{-1} X\hat{\beta} = X^T V^{-1} y,$$

which is

$$\hat{\beta} = (X^T V^{-1} X)^{-1} X^T V^{-1} y,$$

as in (GLS).

The first form for $\hat{u}$ follows from the second equation in (MME). The second follows from this by Corollary 9.3(ii). □

The conditional density of u given y is

$$f(u|y) = f(y, u)/f(y) = f(y|u)f(u)/f(y)$$

(an instance of Bayes's Theorem: see e.g. Haigh (2002), §2.2). We obtain $f(y)$ from $f(y, u)$ by integrating out u (as in §1.5 on the bivariate normal). By above (below (NMM)), $f(y, u)$ is equal to a constant multiplied by

$$\exp\{-\frac{1}{2}[u^T(Z^T R^{-1} Z + D^{-1})u - 2u^T Z^T R^{-1}(y - X\beta) + (y - X\beta)^T R^{-1}(y - X\beta)]\}.$$

This has the form of a multivariate normal. So by Theorem 4.25, $u|y$ is also multivariate normal. We can pick out *which* multivariate normal by identifying the mean and covariance from Edgeworth's Theorem, Theorem 4.16 (see also Note 4.30). Looking at the quadratic term in u above identifies the covariance matrix as $(Z^T R^{-1} Z + D^{-1})^{-1}$. Then looking at the linear term in u identifies the mean as

$$\left(Z^T R^{-1} Z + D^{-1} \right)^{-1} Z^T R^{-1} (y - X\beta).$$

Here β on the right is unknown; replacing it by its BLUP $\hat{\beta}$ gives the first form for $\hat{u}$ (recall from §4.5 that a regression is a conditional mean; this replacement of β by $\hat{\beta}$ is called a *plug-in estimator*). The interpretation of the second form of $\hat{u}$, in terms of the regression of u on y with $\hat{\beta}$ plugged in for β, is similar (as in (GLS), with $(X^T V^{-1} X)^{-1}$ replaced by $(I^T D^{-1} I)^{-1} = D$, X^T by Z^T and y by $y - X\hat{\beta}$.

Note 9.5

1. The use of Bayes's Theorem above is very natural in this context. In Bayesian Statistics, parameters are no longer unknown constants as here. Our initial uncertainty about them is expressed in terms of a distribution, given here by a density, the *prior density*. After sampling and obtaining our data, one uses Bayes's Theorem to update this prior density to a *posterior density*. From this Bayesian point of view, the distinction between fixed and random effects in the mixed model above evaporates. So one can expect simplification, and unification, in a Bayesian treatment of the Linear Model. However, one should first meet a treatment of Bayesian Statistics in general, and for this we must refer the reader elsewhere. For a Bayesian treatment of the Linear Model (fixed effects), see Williams (2001), §8.3.

Bayes's Theorem stems from the work of Thomas Bayes (1702–1761, posthumously in 1764). One of the founders of modern Bayesian Statistics was I. J. Good (1916–2009, from 1950 on). Good also pioneered penalised likelihood, which we met above and will meet again in §9.2 below.

2. In Henderson's mixed model equations (MME), one may combine β and u into one vector, v say, and express (MME) as one matrix equation, $Mv = c$ say. This may be solved as $v = M^{-1}c$. Here, one needs the inverse of the partitioned matrix M. We have encountered this in Exercise 4.10. The relevant Linear Algebra involves the *Schur complement*, and gives an alternative to the approach used above via Woodbury's Formula.

Example 9.6 (Mixed model analysis of ageing athletes)

We give a brief illustration of mixed models with an application to the athletics data in Table 9.1.

In S-Plus/R® the basic command is `lme`, although in R® this requires loading the package `nlme`. We fit a model using Restricted Maximum Likelihood (REML) with fixed effects for the intercept, age and club status, and a random intercept depending on each athlete.

Athlete	Age	Club	Time	Athlete	Age	Club	Time
1	38	0	91.500	4	41	0	91.167
1	39	0	89.383	4	42	0	90.917
1	40	0	93.633	4	43	0	90.883
1	41	0	93.200	4	44	0	92.217
1	42	0	93.533	4	45	1	94.283
1	43	1	92.717	4	46	0	99.100
2	53	1	96.017	5	54	1	105.400
2	54	1	98.733	5	55	1	104.700
2	55	1	98.117	5	56	1	106.383
2	56	1	91.383	5	57	1	106.600
2	58	1	93.167	5	58	1	107.267
2	57	1	88.950	5	59	1	111.133
3	37	1	83.183	6	57	1	90.250
3	38	1	83.500	6	59	1	88.400
3	39	1	83.283	6	60	1	89.450
3	40	1	81.500	6	61	1	96.380
3	41	1	85.233	6	62	1	94.620
3	42	0	82.017				

Table 9.1 Data for Example 9.6. The times are taken from athletes regularly competing in the Berkhamsted Half–Marathon 2002–2007.

```
m1.nlme<-lme(log(time)~club+log(age), random=~1|athlete)
summary(m1.nlme)
```

From the output, t-statistics show that the fixed effects term for age is significant ($p = 0.045$) but suggest that a fixed effects term for club status is not needed ($p = 0.708$). We repeat the analysis, excluding the fixed effects term for club status:

```
m2.nlme<-lme(log(time)~log(age), random=~1|athlete)
```

Next we fit a model with a fixed effect term for age, but allow for the possibility that this coefficient can vary randomly between athletes:

```
m3.nlme<-lme(log(time)~log(age), random=~1+log(age)|athlete)
```

The AIC for these latter two models are -114.883 and -112.378 respectively, so the most appropriate model appears to be the model with a random intercept term and a fixed age-effect term. Log(age) is significant in the chosen model – a t-test gives a p-value of 0.033. A 95% confidence interval for the coefficient of log(age) is 0.229 ± 0.209, consistent with earlier estimates in Examples 3.37 and 8.3, although this time this estimate has a higher level of uncertainty attached to it.

One reason why the ageing effect appears to be weaker here is that the Berkhamsted Half-Marathon (in March) is often used as a 'sharpener' for the London Marathon in April. One could allow for this by using a Boolean variable for London Marathon status (though full data here would be hard to obtain for any data set big enough for the effort to be worthwhile).

9.2 Non-parametric regression

In §4.1 on polynomial regression, we addressed the question of fitting a function $f(x)$ more general than a straight line through the data points in the least-squares sense. Because polynomials of high degree are badly behaved numerically, we restricted attention there to polynomials of low degree. This is a typical parametric setting.

However, we may need to go beyond this rather restricted setting, and if we do the number of parameters we use can increase. This provides more flexibility in fitting. We shall see below how spline functions are useful in this context. But the point here is that we can now move to a function-space setting, where the dimensionality of the function space is infinite. We will use only finitely many parameters. Nevertheless, because the number of parameters available is infinite, and because one usually uses the term *non-parametric* to describe situations with infinitely many parameters, this area is referred to as *non-parametric regression*.

The idea is to choose some suitable set of basic, or simple, functions, and then represent functions as finite linear combinations of these. We have met this before in §4.1, where the basic functions are powers, and §4.1.2, where they are orthogonal polynomials. The student will also have met such ideas in Fourier analysis, where we represent functions as series of sines and cosines (infinite series in theory, finite series in practice). Many other sets of basic functions are in common use – splines, to which we now turn, radial basis functions, wavelets, etc. The relevant area here is Approximation Theory, and we must refer to a text in that area for details and background; see e.g. Ruppert, Wand and Carroll (2003).

The above deals with functions of one variable, or problems with one covariate, but in Chapter 3 we already have extensive experience of problems with several covariates. A similar extension of the treatment to higher dimensions is possible here too. For brevity, we will confine such extensions to two dimensions. Non-parametric regression in two dimensions is important in Spatial Statistics, to which we return in the next subsection.

Recall that in §4.1 on polynomial regression we found that polynomials of high degree are numerically unstable. So if a polynomial of low degree does not suffice, one needs functions of some other kind, and a suitable function class is provided by *splines*. A spline of *degree p* is a continuous function f that is piecewise polynomial of degree p, that is, polynomial of degree p on subintervals $[x_i, x_{i+1}]$, where f and its derivatives $f', \ldots, f^{(p-1)}$ are continuous at the points x_i, called the *knots* of the spline. Typical splines are of the form

$$(x-a)_+^k, \qquad x_+^k := \left\{ \begin{array}{ll} x^k, & x \geq 0, \\ 0, & x < 0. \end{array} \right.$$

We shall restrict ourselves here to *cubic splines*, with $p = 3$; here f, f' and f'' are continuous across the knots x_i. These may be formed by linear combinations of functions of the above type, with $k \leq 3$ and a the knots x_i. It is possible and convenient, to restrict to *basic splines*, or *B-splines*. These are of local character, which is convenient numerically, and one can represent any spline as a linear combination of B-splines. For background and details, see e.g. de Boor (1978).

Suppose now we wish to approximate data y_i at points x_i. As with polynomial regression, we can approximate arbitrarily closely in the least-squares sense, but this is no use to us as the approximating functions are unsuitable. This is because they oscillate too wildly, or are insufficiently smooth. To control this, we need to *penalise* functions that are too rough. It turns out that a suitable measure of roughness for cubic splines is provided by the integral $\int (f'')^2$ of the squared second derivative. We are led to the minimisation problem

$$\min \quad \sum_{i=1}^n (y_i - f(x_i))^2 + \lambda^2 \int (f''(x))^2 \, dx.$$

Here the first term is the sum of squares as before, the integral term is a *roughness penalty*, and λ^2 is called a *smoothing parameter*. (As the sum is of the same form as in the likelihood theory of earlier chapters, and the integral is a penalty term, the method here is called *penalised likelihood* or *penalised log-likelihood*.) With λ small, the roughness penalty is small and the minimiser is close to the least-squares solution as before; with λ large, the roughness penalty is large, and the minimiser will be smooth, at the expense of giving a worse least-squares fit. Since λ is under our control, we have a choice as to how much smoothness we wish, and at what cost in goodness of fit.

It turns out that the minimising function f above is necessarily a cubic spline with knots at the points x_i. This will be a linear combination of B-splines $B_j(x)$, with coefficients β_j say. Forming the β_j into a vector β also, the approximating f is then

$$f(x) = \beta^T B(x),$$

and the mimimisation problem is of the form

$$\min \quad \sum_{i=1}^n \left(y_i - \beta^T B(x_i) \right)^2 + \lambda^2 \beta^T D \beta,$$

for some symmetric positive semi-definite matrix D whose entries are integrals of products of derivatives of the basic splines.

This minimisation problem is of the same form as that in §9.1 for BLUPS, and may be solved in the same way: smoothing splines are BLUPs. Let X be the matrix with ith row $B(x_i)^T$. One obtains the minimising β and fitted values $\hat{y}$ as

$$\hat{\beta} = (X^T X + \lambda^2 D)^{-1} X^T y, \qquad \hat{y} = X(X^T X + \lambda^2 D)^{-1} X^T y = S_\lambda y,$$

say, where S_λ is called the *smoother matrix*. Use of smoothing splines can be implemented in S-Plus/R® by the command `smooth.spline`; see Venables and Ripley (2002), §8.7. For background and details, see Green and Silverman (1994), Ruppert, Wand and Carroll (2003).

Splines were studied by I. J. Schoenberg (1903–1990) from 1946 on, and were used in Statistics by Grace Wahba (1934–) from 1970 on. The term spline derives from the flexible metal strips used by draughtsmen to construct smooth curves interpolating fixed points, in the days before computer-aided design (CAD). Penalised likelihood and roughness penalties go back to I. J. Good (with his student R. A. Gaskins) in 1971 (preceding the AIC in 1974).

9.2.1 Kriging

Kriging describes a technique for non-parametric regression in spatial problems in multiple (commonly three) dimensions. The original motivation was to model ore deposits in mining, though applications extend beyond geology and also typically include remote sensing and black-box modelling of computer experiments. The name kriging derives from the South African mining engineer D. G. Krige (1919–), and was further developed in the 1960s by the French mathematician G. Matheron (1930–2000) at the Paris School of Mines. The basic idea behind kriging is as follows. We observe data

$$(\mathbf{x_1}, y_1), \ldots, (\mathbf{x_n}, y_n),$$

where the $\mathbf{x_i} \in \mathbb{R}^d$ and the $y_i \in \mathbb{R}$. We might imagine the $\mathbf{x_i}$ as a sequence of co-ordinates and the y_i as corresponding to observed levels of mineral deposits. If $d = 2$, this picture corresponds to a three-dimensional plot in which y is the height. Given the observed sequence of $(\mathbf{x_i}, y_i)$ we wish to estimate the y values corresponding to a new set of data $\mathbf{x_0}$. We might, for example, envisage this set-up corresponding to predicting the levels of oil or mineral deposits, or some environmental pollutant etc., at a set of new locations given a set of historical measurements. The set-up for our basic kriging model is

$$y_i = \mu + S(\mathbf{x_i}) + \epsilon_i,$$

where $S(\mathbf{x})$ is a zero-mean stationary stochastic process in $\mathbb{R}^d$ with covariance matrix $\mathbf{C}$ independent of the ϵ_i, which are assumed iid $N(0, \sigma^2)$. However, this formulation can be made more general by choosing $\mu = \mu(x)$ (Venables and Ripley (2002), Ch. 15). It is usually assumed that

$$C_{ij} = \text{cov}\,(S(\mathbf{x_i}, \mathbf{x_j})) = C(||\mathbf{x_i} - \mathbf{x_j}||), \qquad \text{(Isotropy)}$$

although more general models which do not make this assumption are possible. Suppose that the ϵ_i and $S(\cdot)$ are multivariate normal. By §4.6 the mean square error is minimised by the Conditional Mean Formula given by Theorem 4.25. We have that

$$\left(\begin{array}{c} \mathbf{y(x_0)} \\ y(\mathbf{x_0}) \end{array} \right) \sim N\left(\left(\begin{array}{c} \mu\mathbf{1} \\ \mu \end{array} \right), \left(\begin{array}{cc} (\mathbf{C} + \sigma^2 I) & c_0 \\ c_0^T & \sigma^2 \end{array} \right) \right),$$

where $\mathbf{1}$ denotes a column vector of 1s. It follows that the optimal prediction (best linear predictor) for the unobserved $y(\mathbf{x_0})$ given the observed $\mathbf{y(x_0)}$ is given by

$$\hat{y}(\mathbf{x_0}) = \mu + c_0^T \left(\mathbf{C} + \sigma^2 I \right)^{-1} (\mathbf{y(x_0)} - \mu\mathbf{1}). \qquad (BLP)$$

From first principles, it can be shown that this still gives the *best linear predictor (BLP)* when we no longer assume that $S(\mathbf{x})$ and ϵ_i are Gaussian. In practice $\mathbf{C}$ can be estimated using either maximum likelihood or variogram methods (some details can be found in Ruppert, Wand and Carroll (2003), Ch. 13 or Venables and Ripley (2002), Ch. 15). As presented in Ruppert, Wand and Carroll (2003) the full kriging algorithm is as follows:

1. Estimate the covariance function C, σ^2 and set $\mu = \overline{y}$.

2. Construct the estimated covariance matrix $\hat{\mathbf{C}} = C(||x_i - x_j||)$.

3. Set up a mesh of $\mathbf{x_0}$ values in the region of interest.

4. Using (BLP) construct a set of predicted values $\hat{y}(\mathbf{x_0})$.

5. Plot $\hat{y}(\mathbf{x_0})$ against x_0 to estimate the relevant spatial surface.

As briefly discussed in Ruppert, Wand and Carroll (2003), Ch. 13.3–4. it is possible to relate kriging to the non-parametric regression models with a non-parametric regression model using cubic splines. In particular, two-dimensional kriging can be shown to be equivalent to minimising

$$\sum_{i=1}^{n} (y_i - f(x_1, x_2))^2 + \lambda \int \int \left(f_{x_1 x_1}^2 + 2 f_{x_1 x_2}^2 + f_{x_2 x_2}^2 \right) \, dx_1 \, dx_2.$$

This gives an integral of the sum of squares of second derivatives to generalise cubic splines; see e.g. Cressie (1993) §3.4.5 for further details.

The end product of a kriging study may well be some computer graphic, perhaps in (a two-dimensional representation of) three dimensions, perhaps in colour, etc. This would be used to assist policy makers in decision taking – e.g. whether or not to drill a new oil well or mine shaft in some location, whether or not to divert traffic, or deny residential planning permission, for environmental reasons, etc. Specialist software is needed for such purposes.

9.3 Experimental Design

9.3.1 Optimality criteria

We have already seen in §7.1 how to identify unusual data points, in terms of their *leverage* and *influence*. For example, Cook's distance D_i is defined by a quadratic form in the information matrix $C = A^T A$ formed from the design matrix A. Before conducting the statistical experiment that leads to our data y, the design matrix A is still at our disposal, and it is worth considering whether we can choose A in some good way, or better still, in some optimal way. This is indeed so, but there are a number of different possible optimality criteria. One criterion in common use is to maximise the determinant of the information matrix C, the determinant $|C|$ serving as a measure of quantity of information (recall from vector algebra that the volume of a parallelepiped with sides three 3-vectors is the determinant of their co-ordinates).

The situation is similar to that in our first course in Statistics, when we discussed estimation of parameters. Here two important measures of quality of an estimator $\hat{\theta}$ of a parameter θ are *bias*, $E\hat{\theta} - \theta$, and *precision*, measured by the inverse of the variance var θ; we can think of this variance as a measure of sampling error, or noise. We want to keep both noise and bias low, but it is pointless to diminish one at the expense of increasing the other. One thus has a *noise–bias tradeoff*, typical in Statistics. To choose how to make this trade–off, one needs some optimality criterion. This is usually done by choosing some *loss function* (or alternatively, some *utility function*). One then *minimises expected loss* (or *maximises expected utility*). This area of Statistics is called Decision Theory.

The situation here is similar. One needs some optimality criterion for the experimental design (there are a number in common use) – maximising the determinant as above corresponds to D-optimality – and seeks to optimise the design with respect to this criterion. For further detail, we must refer to a book on Optimal Experimental Design, for example Atkinson and Donev (1992).

9.3.2 Incomplete designs

In addition to the profoundly mathematical criteria above, there are also more tangible ways in which experimental design can bring benefits to experimenters by reducing the sample size requirements needed in order to perform a full analysis. It is frequently impractical, say in an agricultural experiment, to grow or include every combination of treatment and block. (Recall that in §2.7 every combination of treatment and block occurred *once*, with multiple replications possible in §2.8.)

Rather than admitting defeat and returning to one-way ANOVA (hence confounding treatment effects with block effects) we need some *incomplete design* which nonetheless enables all treatment and block effects to be estimated. The factors of treatment and block need to be *balanced*, meaning that any two treatments occur together in the same block an equal number of times. This leads to a set of designs known as *balanced incomplete block designs (BIBD)*. These designs are usually tabulated, and can even be used in situations where the blocks are of insufficient size to accommodate one whole treatment allocation (provided that the allocation of experimental units is appropriately randomised). For full details and further reference we refer to Montgomery (1991), Ch. 6. Analysis of large experiments using fractions of the permissible factor combinations is also possible in so-called *factorial* experiments using *fractional factorial designs* (see Montgomery (1991) Ch. 9–12).

Example 9.7 (Latin Squares)

We consider briefly the simplest type of incomplete block design. Suppose we have (e.g.) five types of treatment (fertiliser) to apply to five different varieties of wheat on five different types of soil. This simple experiment leads to 125 different factor combinations in total. It is economically important to be able to test

$$H_0 : \text{The treatment (fertiliser) means are all equal,}$$

in such two-factor experiments (variety and soil type) with fewer than 125 readings. We can make do with 25 readings by means of a 5×5 *Latin square* (see Table 9.2). Each cell contains each fertiliser type once, showing that the design is indeed balanced. Given experimental observations, an ANOVA table with *three* factors (Soil type, Variety and Fertiliser) can be constructed by using the general methods of Chapter 2.

	Variety				
Soil Type	1	2	3	4	5
1	1	2	3	4	5
2	5	1	2	3	4
3	4	5	1	2	3
4	3	4	5	1	2
5	2	3	4	5	1

Table 9.2 5×5 Latin square design. Fertiliser allocations by Soil Type and Variety.

Analysis of $n\times n$ Latin squares. We show how to perform a skeleton ANOVA for a $n\times n$ Latin square design. The approach follows the same general outline laid out in Chapter 2, but generalises §2.6–2.7 by including three factors. In effect, we isolate treatment effects by 'blocking' over rows and columns. The model equation can be written as

$$X_{ijk} = \mu + r_i + c_j + t_k + \epsilon_{ijk}, \quad \epsilon_{ijk} \ \text{iid} \ N(0, \sigma^2),$$

for $i, j = 1 \ldots, n$, where $k = k(i,j)$ is the entry in the Latin square in position (i,j) in the matrix. Note $k = 1, \ldots, n$ also. The r_i, c_j, t_k denote row, column and treatment effects respectively and satisfy the usual constraints:

$$\sum_i r_i = \sum_j c_j = \sum_k t_k = 0.$$

Write

$$R_i = i\text{th row total,} \quad X_{i\bullet} = R_i/n = i\text{th row mean,}$$
$$C_j = j\text{th column total,} \quad X_{\bullet j} = C_j/n = j\text{th column mean,}$$

$$T_k = k\text{th treatment total}, \quad X_{(k)} = T_k/n = k\text{th treatment mean},$$
$$G = \text{grand total} = \sum_i \sum_j \sum_k X_{ijk}, \quad \overline{X} = G/n \quad \text{grand mean}.$$

The following algebraic identity can be verified:

$$SS := SSR + SSC + SST + SSE,$$

where

$$SS := \sum_i \sum_j \sum_k \left(X_{ijk} - \overline{X} \right)^2 = \sum_i \sum_j \sum_k X_{ijk}^2 - \frac{G^2}{n^2},$$

$$SSR := n \sum_i \left(X_{i\bullet} - \overline{X} \right)^2 = \frac{1}{n} \sum_i R_i^2 - \frac{G^2}{n^2},$$

$$SSC := n \sum_j \left(X_{\bullet j} - \overline{X} \right)^2 = \frac{1}{n} \sum_j C_j^2 - \frac{G^2}{n^2},$$

$$SST := n \sum_k \left(X_{(k)} - \overline{X} \right)^2 = \frac{1}{n} \sum_k T_k^2 - \frac{G^2}{n^2},$$

$$SSE := \sum_i \sum_j \sum_k \left(X_{ijk} - X_{i\bullet} - X_{\bullet j} - X_{(k)} + 2\overline{X} \right)^2,$$

with $SSE = SS - SSR - SSC - SST$ as before. An Analysis of Variance of this model can be performed as laid out in Table 9.3.

Source	df	SS	Mean Square	F
Rows	$n-1$	SSR	$MSR = \frac{SSR}{n-1}$	MSR/MSE
Columns	$n-1$	SSC	$MSC = \frac{SSC}{n-1}$	MSC/MSE
Treatments	$n-1$	SST	$MST = \frac{SST}{n-1}$	MST/MSE
Residual	$(n-1)(n-2)$	SSE	$MSE = \frac{SSE}{(n-1)(n-2)}$	
Total	n^2-1	SS		

Table 9.3 ANOVA table for $n \times n$ Latin square

Note 9.8

While Experimental Design is a very useful and practical subject, it also uses a lot of interesting pure mathematics. One area important here is projective geometry over finite fields; see e.g. Hirschfeld (1998). Whereas the mathematics here is discrete, as one would expect since matrix theory is involved, important insights can be gained by using a continuous framework, and so analysis rather than algebra; see e.g. Wynn (1994).

Experimental Design is one of a number of areas pioneered by Fisher in his time at Rothamsted in the 1920s, and by his Rothamsted colleague Frank

Yates (1902–1994). Fisher published his book *The Design of Experiments* in 1935.

9.4 Time series

It often happens that data arrive sequentially in time. This may result in measurements being taken at regular intervals – for example, daily temperatures at noon at a certain meteorological recording station, or closing price of a particular stock, as well as such things as monthly trade figures and the like. We may suppose here that time is measured in discrete units, and that the nth reading is X_n. Then the data set $X = (X_n)$ is called a *time series* (TS).

One often finds in time series that high values tend to be followed by high values, or low values by low values. Typically this is the case when the underlying system has some dynamics (probably complicated and unknown) that tends to fluctuate about some mean value, but intermittently undergoes some perturbation away from the mean in some direction, this perturbation showing a marked tendency to persist for some time, rather than quickly die away.

In such cases one has a *persistence of memory* phenomenon; the question is how long does memory persist? Sometimes memory persists indefinitely, and the infinitely *remote past* continues to exert an influence (rather as the magnetism in a rock reflects the conditions when the rock solidified, in a former geological era, or tempered steel locks in its thermal history as a result of the tempering process). But more commonly only the recent past really influences the present. Using p for the number of parameters as usual, we may represent this by a model in which the present value X_t is influenced by the last p values $X_{t-1}, \ldots, X_{t-p}$. The simplest such model is a linear regression model, with these as covariates and X_t as dependent variable. This gives the model equation

$$X_t = \phi_1 X_{t-1} + \ldots + \phi_p X_{t-p} + \epsilon_t. \qquad (AR(p))$$

Here the ϕ_i are the parameters, forming a vector ϕ, and the ϵ_t are independent errors, normally distributed with mean 0 and common variance σ^2. This gives an *autoregressive model* of *order p*, $AR(p)$, so called because the process X is *regressed on itself*.

For simplicity, we centre at means (that is, assume all $EX_t = 0$) and restrict to the case when $X = (X_n)$ is *stationary* (that is, its distribution is invariant under shifts in time). Then the covariance depends only on the time difference – or rather, its modulus, as the covariance is the same for two variables, either way round; similarly for the correlation, on dividing by the variance σ^2. Write this as $\rho(k)$ at lag k:

$$\rho(k) = \rho(-k) = E[X_t X_{t-k}].$$

Multiplying $(AR(p))$ by X_k and taking expectations gives

$$\rho(k) = \phi_1 \rho(k-1) + \ldots + \phi_p \rho(k-p) \qquad (k > 0). \qquad (YW)$$

These are the *Yule–Walker equations* (G. Udny Yule in 1926, Sir Gilbert Walker in 1931). One has a *difference equation* of *order p*, with *characteristic polynomial*

$$\lambda^p - \phi_1 \lambda^{p-1} - \ldots - \phi_p = 0.$$

If λ_1, ..., λ_p are the roots of this polynomial, then the general solution is

$$\rho(k) = c_1 \lambda_1^k + \ldots + c_p \lambda_p^k$$

(if the roots are distinct, with appropriate modification for repeated roots). Since $\rho(.)$ is a correlation, one has $|\rho(k)| \leq 1$ for all k, which forces

$$|\lambda_i| \leq 1 \qquad (i = 1, \ldots, p).$$

One may instead deal with *moving average* processes of *order q*,

$$X_t = \theta_1 \epsilon_{t-1} + \ldots + \theta_q \epsilon_{t-q} + \epsilon_t, \qquad (MA(q))$$

or with a combination,

$$X_t = \phi_1 X_{t-1} + \ldots + \phi_p X_{t-p} + \theta_1 \epsilon_{t-1} + \ldots + \theta_q \epsilon_{t-q} + \epsilon_t. \quad (ARMA(p,q))$$

The class of autoregressive moving average models, or $ARMA(p,q)$ processes, is quite rich and flexible, and is widely used. We refer to e.g. Box and Jenkins (1970), Brockwell and Davis (2002) for details and background.

9.4.1 Cointegration and spurious regression

Integrated processes. One standard technique used to reduce non-stationary time series to the stationary case is to *difference* them repeatedly (one differencing operation replaces X_t by $X_t - X_{t-1}$). If the series of dth differences is stationary but that of $(d-1)$th differences is not, the original series is said to be *integrated* of *order d*; one writes

$$(X_t) \sim I(d).$$

Cointegration. If $(X_t) \sim I(d)$, we say that (X_t) is *cointegrated* with *cointegration vector* α if $(\alpha^T X_t)$ is (integrated of) order less than d.

A simple example of cointegration arises in random walks. Suppose $X_n = \sum_{i=1}^{n} \xi_i$ with the ξ_n iid random variables, and $Y_n = X_n + \epsilon_n$, with the ϵ_n iid errors as above, is a noisy observation of X_n. Then the bivariate process $(X, Y) = (X_n, Y_n)$ is integrated of order 1, with cointegration vector $(1, -1)^T$.

Cointegrated series are series that tend to move together, and commonly occur in economics. These concepts arose in econometrics, in the work of R. F. Engle (1942–) and C. W. J. (Sir Clive) Granger (1934–2009) in 1987. Engle and Granger gave (in 1991) an illustrative example – the prices of tomatoes in North Carolina and South Carolina. These states are close enough for a significant price differential between the two to encourage sellers to transfer tomatoes to the state with currently higher prices to cash in; this movement would increase supply there and reduce it in the other state, so supply and demand would move the prices towards each other.

Engle and Granger received the Nobel Prize in Economics in 2003. The citation included the following:

> Most macroeconomic time series follow a stochastic trend, so that a temporary disturbance in, say, GDP has a long-lasting effect. These time series are called nonstationary; they differ from stationary series which do not grow over time, but fluctuate around a given value. Clive Granger demonstrated that the statistical methods used for stationary time series could yield wholly misleading results when applied to the analysis of nonstationary data. His significant discovery was that specific combinations of nonstationary time series may exhibit stationarity, thereby allowing for correct statistical inference. Granger called this phenomenon cointegration. He developed methods that have become invaluable in systems where short-run dynamics are affected by large random disturbances and long-run dynamics are restricted by economic equilibrium relationships. Examples include the relations between wealth and consumption, exchange rates and price levels, and short and long-term interest rates.

Spurious regression. Standard least-squares models work perfectly well if they are applied to *stationary* time series. But if they are applied to *non-stationary* time series, they can lead to spurious or nonsensical results. One can give examples of two time series that clearly have nothing to do with one another, because they come from quite unrelated contexts, but nevertheless have quite a high value of R^2. This would normally suggest that a correspondingly high proportion of the variability in one is accounted for by variability in the other – while in fact *none* of the variability is accounted for. This is the phenomenon of *spurious regression*, first identified by Yule in 1927, and later studied by

Granger and Newbold in 1974. We can largely avoid such pitfalls by restricting attention to stationary time series, as above.

ARCH and GARCH.

The terms homoscedastic and heteroscedastic are used to describe processes where the variance is constant or is variable. With Z_i independent and normal $N(0,1)$, the *autoregressive conditionally heteroscedastic (ARCH)* model of *order* p, or $ARCH(p)$, is defined by the model equations

$$X_t = \sigma_t Z_t, \qquad \sigma_t^2 = \alpha_0 + \sum_{i=1}^p \alpha_i X_{t-i}^2, \qquad (ARCH(p))$$

for $\alpha_0 > 0$ and $\alpha_i \geq 0$. The $AR(p)$ character is seen on the right of the second equation; the *conditional variance* of X_t *given* the information available at time $t-1$ is σ_t^2, a function of $X_{t-1}, \ldots, X_{t-p}$, and so varies, hence the conditional heteroscedasticity. In the *generalised* ARCH model $GARCH(p,q)$, the variance becomes

$$\sigma_t^2 = \alpha_0 + \sum_{i=1}^p \alpha_i X_{t-i}^2 + \sum_{j=1}^q \beta_j X \sigma_{t-j}^2. \qquad (GARCH(p,q))$$

Both ARCH and GARCH models are widely used in econometrics; see e.g. Engle's Nobel Prize citation. We must refer to a specialist time series or econometrics textbook for more details; the point to note here is that regression methods are widely used in economics and econometrics.

Note 9.9

We observed in §1.2.2 and §7.1 that, while independent errors tend to cancel as in the Law of Large Numbers, strongly dependent errors need not do so and are very dangerous in Statistics. The time series models above, which can model the tendency of high or low values to follow each other, reflect this – though there we separate out the terms giving rise to this and put them in the main part of the model, rather than the error.

9.5 Survival analysis

We return to the Poisson point process, $Ppp(\lambda)$ say, first discussed in §8.4. In the sequel the parameter λ has the interpretation of an *intensity* or *rate* as follows. For an interval I of length $|I|$, the number of points of the process (the number of *Poisson points*) is Poisson distributed with parameter $\lambda|I|$; the counts in disjoint intervals are independent. This use of an intensity parameter to measure exposure to risk (of mortality) is generalised below.

Suppose now we have a population of individuals, whose lifetimes are independent, each with distribution function F on $(0, \infty)$, which we will suppose to have density f. If T is the lifetime of a given individual, the conditional probability of death in a short interval $(t, t + h)$ given survival to time t is, writing $\overline{F}(t) := 1 - F(t) = P(T > t)$ for the tail of F,

$$P(T \in (t, t + h)|T > t) = P(T \in (t + h))/P(T > t) = hf(t)/\overline{F}(t),$$

to first order in h. We call the coefficient of h on the right the *hazard function*, $h(t)$. Thus

$$h(t) = f(t) / \int_t^\infty f(u) \, du = -D\left(\int_t^\infty f\right) / \int_t^\infty f,$$

and integrating one has

$$\log\left(\int_t^\infty f\right) = -\int_0^t h : \qquad \int_t^\infty f(u) \, du = \exp\left\{-\int_0^t h(u) \, du\right\}$$

(since f is a density, $\int_0^\infty f = 1$, giving the constant of integration).

Example 9.10

1. *The exponential distribution.* If F is the exponential distribution with parameter λ, $E(\lambda)$ say, $f(t) = \lambda e^{-\lambda t}$, $\overline{F}(t) = e^{-\lambda t}$, and $h(t) = \lambda$ is constant. This property of constant hazard rate captures the *lack of memory property* of the exponential distributions (for which see e.g. the sources cited in §8.4), or the lack of ageing property: given that an individual has survived to date, its further survival time has the same distribution as that of a new individual. This is suitable for modelling the lifetimes of certain components (lightbulbs, etc.) that fail without warning, but of course not suitable for modelling lifetimes of biological populations, which show ageing.
2. *The Weibull distribution.*
   Here

$$f(t) = \lambda \nu^{-\lambda} t^{\lambda-1} \exp\{-(t/\lambda)^\nu\},$$

with λ, ν positive parameters; this reduces to the exponential $E(\lambda)$ for $\nu = 1$.
3. *The Gompertz-Makeham distribution.*
   This is a three-parameter family, with hazard function

$$h(t) = \lambda + ae^{bt}.$$

This includes the exponential case with $a = b = 0$, and allows one to model a baseline hazard (the constant term λ), with in addition a hazard growing exponentially with time (which can be used to model the winnowing effect of ageing in biological populations).

In medical statistics, one may be studying survival times in patients with a particular illness. One's data is then subject to *censoring*, in which patients may die from other causes, discontinue treatment, leave the area covered by the study, etc.

9.5.1 Proportional hazards

One is often interested in the effect of covariates on survival probabilities. For example, many cancers are age-related, so the patient's age is an obvious co-variate. Many forms of cancer are affected by diet, or lifestyle factors. Thus the link between smoking and lung cancer is now well known, and similarly for exposure to asbestos. One's chances of contracting certain cancers (of the mouth, throat, oesophagus etc.) are affected by alcohol consumption. Breast cancer rates are linked to diet (western women, whose diets are rich in dairy products, are more prone to the disease than oriental women, whose diets are rich in rice and fish). Consumption of red meat is linked to cancer of the bowel, etc., and so is lack of fibre. Thus in studying survival rates for a particular cancer, one may identify a suitable set of covariates z relevant to this cancer. One may seek to use a linear combination $\beta^T z$ of such covariates with coefficients β, as in the multiple regression of Chapters 3 and 4. One might also superimpose this effect on some baseline hazard, modelled non-parametrically. One is led to model the hazard function by

$$h(t; z) = g(\beta^T z)h_0(t),$$

where the function g contains the parametric part $\beta^T z$ and the baseline hazard h_0 the non-parametric part. This is the *Cox proportional hazards model* (D. R. Cox in 1972). The name arises because if one compares the hazards for two individuals with covariates z_1, z_2, one obtains

$$h(t; z_1)/h(t; z_2) = g(\beta^T z_1)/g(\beta^T z_2),$$

as the baseline hazard term cancels.

The most common choices of g are:
(i) *Log-linear*: $g(x) = e^x$ (if $g(x) = e^{ax}$, one can absorb the constant a into β);
(ii) *Linear*: $g(x) = 1 + x$;
(iii) *Logistic*: $g(x) = \log(1 + x)$.

We confine ourselves here to the log-linear case, the commonest and most important. Here the hazard ratio is

$$h(t; z_1)/h(t; z_2) = \exp\left\{\beta^T(z_1 - z_2)\right\}.$$

Estimation of β by maximum likelihood must be done numerically (we omit the non-parametric estimation of h_0). For a sample of n individuals, with covariate vectors $z_1, \ldots, z_n$, the data consist of the point events occurring – the identities (or covariate values) and times of death or censoring of non-surviving individuals; see e.g. Venables and Ripley (2002), §13.3 for use of S-Plus here, and for theoretical background see e.g. Cox and Oakes (1984).

9.6 $p \gg n$

We have constantly emphasised that the number p of parameters is to be kept small, to give an economical description of the data in accordance with the Principle of Parsimony, while the sample size n is much larger – the larger the better, as there is then more information. However, practical problems in areas such as *bioinformatics* have given rise to a new situation, in which this is reversed, and one now has p much larger than n. This happens with, for example, data arising from *microarrays*. Here p is the number of entries in a large array or matrix, and p being large enables many biomolecular probes to be carried out at the same time, so speeding up the experiment. But now new and efficient variable-selection algorithms are needed. Recent developments include that of LASSO (least absolute shrinkage and selection operator) and LARS (least angle regression). One seeks to use such techniques to eliminate most of the parameters, and reduce to a case with $p \ll n$ that can be handled by traditional methods. That is, one seeks systematic ways to take a large and complex problem, in which most of the parameters are unimportant, and focus in on the small subset of important parameters.

Solutions

Chapter 1

1.1

$$\begin{aligned} Q(\lambda) &= \lambda^2 \frac{1}{n}\sum_1^n (x_i - \overline{x})^2 + 2\lambda \frac{1}{n}\sum_1^n (x_i - \overline{x})(y_i - \overline{y}) + \frac{1}{n}\sum_1^n (y_i - \overline{y})^2 \\ &= \lambda^2 \overline{(x - \overline{x})^2} + 2\lambda \overline{(x - \overline{x})(y - \overline{y})} + \overline{(y - \overline{y})^2} \\ &= \lambda^2 S_{xx} + 2\lambda S_{xy} + S_{yy}. \end{aligned}$$

Now $Q(\lambda) \geq 0$ for all λ, so $Q(\cdot)$ is a quadratic which does not change sign. So its discriminant is ≤ 0 (if it were > 0, there would be distinct real roots and a sign change in between). So ('$b^2 - 4ac \leq 0$'):

$$s_{xy}^2 \leq s_{xx} s_{yy} = s_x^2 s_y^2, \quad r^2 := (s_{xy}/s_x s_y)^2 \leq 1.$$

So

$$-1 \leq r \leq +1,$$

as required.

The extremal cases $r = \pm 1$, or $r^2 = 1$, have discriminant 0, that is $Q(\lambda)$ has a repeated real root, λ_0 say. But then $Q(\lambda_0)$ is the sum of squares of $\lambda_0(x_i - \overline{x}) + (y_i - \overline{y})$, which is zero. So each term is 0:

$$\lambda_0(x_i - \overline{x}) + (y_i - \overline{y}) = 0 \quad (i = 1, \ldots, n).$$

That is, all the points (x_i, y_i) $(i = 1, \ldots, n)$, lie on a straight line through the centroid $(\overline{x}, \overline{y})$ with slope $-\lambda_0$.

1.2

Similarly

$$
\begin{aligned}
Q(\lambda) &= E\left[\lambda^2 (x - Ex)^2 + 2\lambda(x - Ex)(y - Ey) + (y - Ey)^2\right] \\
&= \lambda^2 E[(x - Ex)^2] + 2\lambda E[(x - Ex)(y - Ey)] + E\left[(y - Ey)^2\right] \\
&= \lambda^2 \sigma_x^2 + 2\lambda \sigma_{xy} + \sigma_y^2.
\end{aligned}
$$

(i) As before $Q(\lambda) \geq 0$ for all λ, as the discriminant is ≤ 0, i.e.

$$
\sigma_{xy}^2 \leq \sigma_x^2 \sigma_y^2, \quad \rho := (\sigma_{xy}/\sigma_x \sigma_y)^2 \leq 1, \quad -1 \leq \rho \leq +1.
$$

The extreme cases $\rho = \pm 1$ occur iff $Q(\lambda)$ has a repeated real root λ_0. Then

$$
Q(\lambda_0) = E[(\lambda_0(x - Ex) + (y - Ey))^2] = 0.
$$

So the random variable $\lambda_0(x - Ex) + (y - Ey)$ is zero (a.s. – except possibly on some set of probability 0). So all values of (x, y) lie on a straight line through the centroid (Ex, Ey) of slope $-\lambda_0$, a.s.

1.3

(i) Half-marathon: $a = 3.310$ (2.656, 3.964). $b = 0.296$ (0.132, 0.460).
Marathon: $a = 3.690$ (2.990, 4.396). $b = 0.378$ (0.202, 0.554).
(ii) Compare rule with model $y = e^a t^b$ and consider, for example, $\frac{dy}{dt}(\bar{t})$. Should obtain a reasonable level of agreement.

1.4

A plot gives little evidence of curvature and there does not seem to be much added benefit in fitting the quadratic term. Testing the hypothesis $c = 0$ gives a p-value of 0.675. The predicted values are 134.44 and 163.89 for the linear model and 131.15 and 161.42 for the quadratic model.

1.5

The condition in the text becomes

$$
\begin{pmatrix} S_{uu} & S_{uv} \\ S_{uv} & S_{vv} \end{pmatrix} \begin{pmatrix} a \\ b \end{pmatrix} = \begin{pmatrix} S_{yu} \\ S_{yv} \end{pmatrix}.
$$

We can write down the solution for $(a\ b)^T$ as

$$
\begin{pmatrix} S_{uu} & S_{uv} \\ S_{uv} & S_{vv} \end{pmatrix}^{-1} \begin{pmatrix} S_{yu} \\ S_{yv} \end{pmatrix} = \frac{1}{S_{uu}S_{vv} - S_{uv}^2} \begin{pmatrix} S_{vv} & -S_{uv} \\ -S_{uv} & S_{uu} \end{pmatrix} \begin{pmatrix} S_{yu} \\ S_{yv} \end{pmatrix},
$$

giving

$$
a = \frac{S_{vv}S_{yu} - S_{uv}S_{yv}}{S_{uu}S_{vv} - S_{uv}^2}, \quad b = \frac{S_{uu}S_{yv} - S_{uv}S_{yu}}{S_{uu}S_{vv} - S_{uv}^2}.
$$

1.6

(i) A simple plot suggests that a quadratic model might fit the data well (leaving aside, for the moment, the question of interpretation). An increase in R^2, equivalently a large reduction in the residual sum of squares, suggests the quadratic model offers a meaningful improvement over the simple model $y = a + bx$. A t-test for $c = 0$ gives a p-value of 0.007.

(ii) t-tests give p-values of 0.001 (in both cases) that b and c are equal to zero. The model has an R^2 of 0.68, suggesting that this simple model explains a reasonable amount, around 70%, of the variability in the data. The estimate gives $c = -7.673$, suggesting that club membership has improved the half-marathon times by around seven and a half minutes.

1.7

(i) The residual sums of squares are 0.463 and 0.852, suggesting that the linear regression model is more appropriate.

(ii) A t-test gives a p-value of 0.647, suggesting that the quadratic term is not needed. (Note also the very small number of observations.)

1.8

A simple plot suggests a faster-than-linear growth in population. Sensible suggestions are fitting an exponential model using $\log(y) = a + bt$, or a quadratic model $y = a + bt + ct^2$. A simple plot of the resulting fits suggests the quadratic model is better, with all the terms in this model highly significant.

1.9

(i) Without loss of generality assume $g(\cdot)$ is a monotone increasing function. We have that $F_Y(x) = \mathrm{P}(g(X) \leq x) = \mathrm{P}(X \leq g^{-1}(x))$. It follows that

$$
\begin{aligned}
f_Y(x) &= \frac{d}{dx} \int_{-\infty}^{g^{-1}(x)} f_X(u)\, du, \\
&= f_X\left(g^{-1}(x)\right) \left(\frac{dg^{-1}(x)}{dx}\right).
\end{aligned}
$$

(ii)

$$
\begin{aligned}
\mathrm{P}(Y \leq x) &= \mathrm{P}(e^X \leq x) = \mathrm{P}(X \leq \log x), \\
f_Y(x) &= \frac{d}{dx} \int_{\infty}^{\log x} \frac{1}{\sqrt{2\pi}\sigma} e^{-\frac{(y-\mu)^2}{2\sigma^2}}\, dy, \\
&= \frac{1}{\sqrt{2\pi}\sigma} x^{-1} \exp\left\{-\frac{(\log x - \mu)^2}{2\sigma^2}\right\}.
\end{aligned}
$$

1.10

(i) $\mathrm{P}\left(Y \leq x\right) = \mathrm{P}\left(r/U \leq x\right) = \mathrm{P}\left(U \geq r/x\right)$. We have that

$$
\begin{aligned}
f_Y(x) &= \frac{d}{dx} \int_{r/x}^{\infty} \frac{\left(\frac{1}{2}\right)^{\frac{r}{2}} u^{\frac{r}{2}-1} e^{-\frac{u}{2}} \, du}{\Gamma(\frac{r}{2})}, \\
&= \frac{\left(\frac{r}{x^2}\right)\left(\frac{1}{2}\right)^{\frac{r}{2}} \left(\frac{r}{x}\right)^{\frac{r}{2}-1} e^{-\frac{r}{2x}}}{\Gamma\left(\frac{r}{2}\right)}, \\
&= \frac{r^{\frac{r}{2}} x^{-1-\frac{r}{2}} e^{-\frac{r}{2x}}}{2^{\frac{r}{2}} \Gamma\left(\frac{r}{2}\right)}.
\end{aligned}
$$

(ii) $\mathrm{P}(Y \leq x) = \mathrm{P}(X \geq 1/x)$, and this gives

$$
\begin{aligned}
f_Y(x) &= \frac{d}{dx} \int_{\frac{1}{x}}^{\infty} \frac{u^{a-1} b^a e^{-bu} \, du}{\Gamma(a)}, \\
&= \frac{\left(\frac{1}{x^2}\right) b^a \left(\frac{1}{x}\right)^{a-1} e^{-b/x}}{\Gamma(a)}, \\
&= \frac{b^a x^{-1-a} e^{-b/x}}{\Gamma(a)}.
\end{aligned}
$$

Since the above expression is a probability density, and therefore integrates to one, this gives

$$
\int_0^{\infty} x^{-1-a} e^{-b/x} \, dx = \frac{\Gamma(a)}{b^a}.
$$

1.11

We have that $f(x,u) = f_Y(u)\phi(x|0,u)$ and $f_{t(r)}(x) = \int_0^{\infty} f(x,u)du$, where $\phi(\cdot)$ denotes the probability density of $N(0,u)$. Writing this out explicitly gives

$$
\begin{aligned}
f_{t_r}(x) &= \int_0^{\infty} \frac{r^{\frac{r}{2}} u^{-1-\frac{r}{2}} e^{-\frac{r}{2u}}}{2^{\frac{r}{2}} \Gamma\left(\frac{r}{2}\right)} \cdot \frac{e^{-\frac{x^2}{2u}}}{\sqrt{2\pi} u^{\frac{1}{2}}} \, du, \\
&= \frac{r^{\frac{r}{2}}}{2^{\frac{r}{2}} \Gamma(\frac{r}{2})\sqrt{2\pi}} \int_0^{\infty} u^{-\frac{3}{2}-\frac{r}{2}} e^{-\left[\frac{r}{2}+\frac{x^2}{2}\right]\frac{1}{u}} \, du, \\
&= \frac{r^{\frac{r}{2}}}{2^{\frac{r}{2}} \Gamma(\frac{r}{2})\sqrt{2\pi}} \frac{\Gamma\left(\frac{r}{2}+\frac{1}{2}\right)}{\left[\frac{r}{2}+\frac{x^2}{2}\right]^{\left(\frac{1}{2}+\frac{r}{2}\right)}}, \\
&= \frac{\Gamma\left(\frac{r}{2}+\frac{1}{2}\right)}{\sqrt{\pi r}\Gamma\left(\frac{r}{2}\right)} \left(1+\frac{x^2}{r}\right)^{-\frac{1}{2}(r+1)}.
\end{aligned}
$$

Chapter 2

2.1

(i)

$$
\begin{aligned}
\int_0^z h(u)\, du &= P(Z \le z) = P(X/Y \le z) = \int\int_{x/y \le z} f(x,y)\, dx\, dy \\
&= \int_0^\infty dy \int_0^{yz} dx\, f(x,y).
\end{aligned}
$$

Differentiate both sides w.r.t. z:

$$
h(z) = \int_0^\infty dy\; y f(yz, y) \quad (z > 0),
$$

as required (assuming enough smoothness to differentiate under the integral sign, as we do here).

(ii) $\int_0^x f_{X/c}(u)\, du = P(X/c \le x) = P(X \le cx) = \int_0^{cx} f_X(u)\, du$. Differentiate w.r.t. x:

$$
f_{X/c}(x) = c f_X(cx),
$$

as required.

(iii) As $\chi^2(n)$ has density

$$
\frac{e^{-\frac{1}{2}x} x^{\frac{1}{2}n - 1}}{2^{\frac{1}{2}n} \Gamma(\frac{1}{2}n)},
$$

χ^2/n has density, by (ii),

$$
\frac{n e^{-\frac{1}{2}nx}(nx)^{\frac{1}{2}n - 1}}{2^{\frac{1}{2}n} \Gamma(\frac{1}{2}n)} = \frac{n^{\frac{1}{2}n} e^{-\frac{1}{2}nx} x^{\frac{1}{2}n - 1}}{2^{\frac{1}{2}n} \Gamma(\frac{1}{2}n)}.
$$

So $F(m,n) := \frac{\chi^2(m)/m}{\chi^2(n)/n}$ (independent quotient) has density, by (i),

$$
\begin{aligned}
h(z) &= \int_0^\infty y \cdot \frac{m^{\frac{1}{2}m}}{2^{\frac{1}{2}m}\Gamma(\frac{1}{2}m)} e^{-\frac{1}{2}myz} z^{\frac{1}{2}m-1} y^{\frac{1}{2}m-1} \frac{n^{\frac{1}{2}n}}{2^{\frac{1}{2}n}\Gamma(\frac{1}{2}n)} e^{-\frac{1}{2}ny} y^{\frac{1}{2}n-1}\, dy \\
&= \frac{m^{\frac{1}{2}m} n^{\frac{1}{2}n} z^{\frac{1}{2}m-1}}{2^{\frac{1}{2}(m+n)} \Gamma(\frac{1}{2}m)\Gamma(\frac{1}{2}n)} \int_0^\infty e^{-\frac{1}{2}(n+mz)y} y^{\frac{1}{2}(m+n)-1}\, dy.
\end{aligned}
$$

Put $\frac{1}{2}(n + mz)y = u$ in the integral, which becomes

$$
\frac{1}{(\frac{1}{2})^{(m+n)}} \int_0^\infty \frac{e^{-u} u^{\frac{1}{2}(m+n)-1}}{(n+mz)^{\frac{1}{2}(m+n)}}\, du = \frac{\Gamma(\frac{1}{2}(m+n))}{(\frac{1}{2})^{(m+n)} (n+mz)^{\frac{1}{2}(m+n)}}.
$$

Combining,

$$h(z) = m^{\frac{1}{2}m} n^{\frac{1}{2}n} \frac{\Gamma(\frac{1}{2}(m+n))}{\Gamma(\frac{1}{2}m)\Gamma(\frac{1}{2}n)} \frac{z^{\frac{1}{2}m-1}}{(n+mz)^{\frac{1}{2}(m+n)}},$$

as required.

2.2
(i) 0.726. (ii) 0.332. (iii) 0.861. (iv) 0.122. (v) 0.967.

2.3
The ANOVA table obtained is shown in Table 1. The significant p-value obtained ($p = 0.007$) gives strong evidence that the absorption levels vary between the different types of fats. The mean levels of fat absorbed are Fat 1 172g, Fat 2 185g, Fat 3 176g, Fat 4 162g. There is some suggestion that doughnuts absorb relatively high amounts of Fat 2, and relatively small amounts of Fat 4.

Source	df	Sum of Squares	Mean Square	F
Between fats	3	1636.5	545.5	5.406
Residual	20	2018.0	100.9	
Total	23	3654.5		

Table 1   One-way ANOVA table for Exercise 2.3

2.4
The one-way ANOVA table is shown in Table 2. The p-value obtained, $p = 0.255$, suggests that the length of daily light exposure does not affect growth.

Source	df	Sum of Squares	Mean Square	F
Photoperiod	3	7.125	2.375	1.462
Residual	20	32.5	1.625	
Total	23	39.625		

Table 2   One-way ANOVA table for Exercise 2.4

2.5
(i) The statistic becomes

$$t = \frac{\sqrt{n}(\overline{X}_1 - \overline{X}_2)}{\sqrt{2}s},$$

where s^2 is the pooled variance estimate given by

$$s^2 = \left[\frac{s_1^2 + s_2^2}{2}\right].$$

(ii) The total sum of squares SS can be calculated as

$$\sum X_{ij}^2 - \frac{n}{2}\left(\overline{X}_1 + \overline{X}_2\right)^2 = \sum X_1^2 + \sum X_2^2 - \frac{n}{2}\left(\overline{X}_1^2 + 2\overline{X}_1\overline{X}_2 + \overline{X}_2^2\right).$$

Similarly,

$$SSE = \left(\sum X_1^2 - n\overline{X}_1^2\right) + \left(\sum X_2^2 - n\overline{X}_2^2\right).$$

This leaves the treatments sum of squares to be calculated as

$$SST = \frac{n}{2}\left(\overline{X}_1^2 - 2\overline{X}_1\overline{X}_2 + \overline{X}_2^2\right) = \frac{n}{2}\left(\overline{X}_1 - \overline{X}_2\right)^2,$$

on 1 degree of freedom since there are two treatments. Further, since by subtraction we have $2(n-1)$ residual degrees of freedom, the F statistic can be constructed as

$$F = \frac{\frac{\frac{n}{2}\left(\overline{X}_1 - \overline{X}_2\right)^2}{1}}{\frac{2(n-1)s^2}{2(n-1)}} = \frac{n\left(\overline{X}_1 - \overline{X}_2\right)^2}{2s^2},$$

and can be tested against $F_{1,2(n-1)}$. We see from (i) that F is the square of the usual t statistic.

2.6
By definition $Y_1^2 + Y_2^2 \sim \chi_2^2$. Set

$$a\left(Y_1 - Y_2\right)^2 + b\left(Y_1 + Y_2\right)^2 = Y_1^2 + Y_2^2.$$

It follows that $aY_1^2 + bY_1^2 = Y_1^2$, $aY_2^2 + bY_2^2 = Y_2^2$, $-2aY_1Y_2 + 2bY_1Y_2 = 0$. Hence $a = b = 1/2$.

2.7
By Theorem 2.4

$$\left[Y_1^2 + Y_2^2 + Y_3^2 - \frac{(Y_1 + Y_2 + Y_3)^2}{3}\right] \sim \chi_2^2.$$

The result follows since the LHS can be written as

$$\frac{1}{3}\left[2Y_1^2 + 2Y_2^2 + 2Y_3^2 - 2\left(Y_1Y_2 + Y_1Y_3 + Y_2Y_3\right)\right],$$

or equivalently as

$$\frac{1}{3}\left[(Y_1 - Y_2)^2 + (Y_2 - Y_3)^2 + (Y_2 - Y_3)^2\right].$$

Continuing, we may again apply Theorem 2.4 to obtain

$$\left[\sum_{i=1}^{n} Y_i^2 - \frac{\left(\sum_{i=1}^{n} Y_i\right)^2}{n}\right] \sim \chi_{n-1}^2.$$

The LHS can be written as

$$\left[\frac{n-1}{n}\sum_{i=1}^{n} Y_i^2 - \frac{2}{n}\sum_{i<j} Y_i Y_j\right] = \frac{1}{n}\left[\sum_{i<j}(Y_i - Y_j)^2\right],$$

and the result generalises as

$$\frac{1}{n}\left[\sum_{i<j}(Y_i - Y_j)^2\right] \sim \chi_{n-1}^2.$$

2.8

The two-way ANOVA table is shown in Table 3. We have significant evidence for differences between the five treatments. The mean numbers of failures for each treatment are: Check 10.8, Arasan 6.2, Spergon 8.2, Semesan Jr. 6.6, Fermate 5.8. It appears that we have greater numbers of failures under the Check and Spergon treatments, with the remaining treatments approximately equally effective. The p-value for the replications term is borderline significant. However, the F-ratio is greater than 2, a result that is usually taken to mean that replication is successful in reducing the residual sum of squares and improving the precision with which treatment effects can be estimated.

Source	df	Sum of Squares	Mean Square	F	p
Treatment	4	83.84	20.96	3.874	0.022
Replication	4	49.84	12.46	2.303	0.103
Residual	16	86.56	5.41		
Total	24	220.24			

Table 3 Two-way ANOVA table for Exercise 2.8

2.9

The two-way ANOVA is shown in Table 4. Note that in the light of additional data both Photoperiod and Genotype are seen to be highly significant. With reference to Exercise 2.4 Photoperiod is important, but only once the effects of different Genotypes are accounted for. Exercises 2.4 and 2.9 nicely illustrate the principles of blocking. Variation due to differing Genotypes is identified and removed from the residual sum of squares in Table 2. When the F ratio is calculated using this much smaller residual mean square, the Photoperiod term is clearly seen to be significant.

Source	df	Sum of Squares	Mean Square	F	p
Photoperiod	3	7.125	2.375	7.703	0.002
Genotype	5	27.875	5.575	18.081	0.000
Residual	15	4.625	0.308		
Total	23	39.625			

Table 4 Two-way ANOVA table for Exercise 2.9

2.10
The two-way ANOVA table is shown in Table 5. Both variety and location appear highly significant, but the interaction term is non-significant. A two-way ANOVA table without interactions is shown in Table 6. Here, both the variety and location terms remain highly significant and there is a sense in which conclusions are resistant to the inclusion of interaction terms. The mean yields for each variety are: A 12.17 bushels, B 17.83 bushels, C 15.67 bushels. In conclusion, both variety and location affect yield. Variety B appears to give the highest yield.

Source	df	Sum of Squares	Mean Square	F	p
Variety	2	196.22	98.11	9.150	0.001
Location	3	468.22	156.07	14.556	0.000
Location:Variety	6	78.406	13.068	1.219	0.331
Residual	23	257.33	10.72		
Total	35	1000.22			

Table 5 Two-way ANOVA table with interactions for Exercise 2.10

2.11
The two-way ANOVA table with interactions is shown in Table 7. We have strong evidence for differences between high and low levels of protein. High protein diets appear to lead to larger weight gains. We have no evidence for differences between the different sources of protein, although a borderline significant p-value gives at least some suggestion that there may be an interaction

Source	df	Sum of Squares	Mean Square	F	p
Variety	2	196.22	98.11	9.150	0.001
Location	3	468.22	156.07	13.944	0.000
Residual	30	335.78	11.19		
Total	35	1000.22			

Table 6 Two-way ANOVA table, restricted to main effects, for Exercise 2.10

between level and source of protein. Tabulated mean values per cell are shown in Table 8 and these appear to reinforce the earlier interpretation of higher weight gains under the high protein diets.

Source	df	Sum of Squares	Mean Square	F	p
Protein level	1	3168.3	3168.3	14.767	0.000
Protein source	2	266.5	133.3	0.621	0.541
Level:source	2	1178.1	589.1	2.746	0.073
Residual	54	11586.0	214.6		
Total	59	16198.93			

Table 7 Two-way ANOVA table with interactions for Exercise 2.11

Source	High Protein	Low Protein
Beef	100	79.2
Cereal	85.9	83.9
Pork	99.5	78.7

Table 8 Mean weight gains per cell for Exercise 2.11

Chapter 3

3.2

$$A = \begin{pmatrix} 1 & 0 & 0 & -1 \\ -1 & 0 & 1 & 0 \\ -1 & 1 & 0 & 0 \\ 0 & -1 & 1 & 0 \\ 0 & 0 & 1 & -1 \\ 0 & 0 & 0 & 1 \end{pmatrix}, \quad C = A^T A = \begin{pmatrix} 3 & -1 & -1 & -1 \\ -1 & 2 & -1 & 0 \\ -1 & -1 & 3 & -1 \\ -1 & 0 & -1 & 3 \end{pmatrix}.$$

$$C^{-1} = \frac{1}{8} \begin{pmatrix} 13 & 12 & 11 & 8 \\ 12 & 16 & 12 & 8 \\ 11 & 12 & 13 & 8 \\ 8 & 8 & 12 & 8 \end{pmatrix}, \quad C^{-1}A^T = \frac{1}{8} \begin{pmatrix} 5 & -2 & -1 & -1 & 3 & 8 \\ 4 & 0 & 4 & -4 & 4 & 8 \\ 3 & 2 & 1 & 1 & 5 & 8 \\ 0 & 0 & 0 & 0 & 0 & 8 \end{pmatrix}.$$

$$P = AC^{-1}A^T = \frac{1}{8} \begin{pmatrix} 5 & -2 & -1 & -1 & 3 & 0 \\ -2 & 4 & 2 & 2 & 2 & 0 \\ -1 & 2 & 5 & -3 & 1 & 0 \\ -1 & 2 & -3 & 5 & 1 & 0 \\ 3 & 2 & 1 & 1 & 5 & 0 \\ 0 & 0 & 0 & 0 & 0 & 8 \end{pmatrix},$$

$$\hat{\beta} = C^{-1}A^T y = \begin{pmatrix} 58.988 \\ 79.56 \\ 83.243 \\ 46.81 \end{pmatrix}.$$

Fitted values can be calculated as

$$AC^{-1}A^T y = (12.178, 24.255, 20.573, 3.683, 36.433, 46.81)^T.$$

Continuing, $SSE = 0.08445$, giving

$$\hat{\sigma} = \sqrt{\frac{SSE}{n-p}} = \sqrt{\frac{0.08445}{6-4}} = 0.205 \text{m}.$$

3.3
(i) Set $R_{1,a}^2 < R_{2,a}^2$.
(ii) Replace $(n-1-p)$ with $(n-j-p)$.

3.4

(i) A plot of Y against Z suggests that some kind of curvature in the relationship between Y and Z might be appropriate. However, once we superimpose the fitted values of the simple model $Y = a + bZ$, it becomes clear that this apparent curvature may simply be as a result of non-constant variance. Rather than considering exotic model formulae with high-order terms in Z we are led to consider the transformation methods of Chapter 7 (See Exercise 7.10).

(ii) Fitting this model, it does not appear that the quadratic term in Z is necessary (see computer output in Table 9). A natural next step is to fit the model $Y = a + bX + cZ$. In this case t-statistics suggest that X is not needed ($p = 0.134$) and we are left with the simple model $Y = a + bZ$. Again, see Exercise 7.10 for a re-analysis in the light of later theory.

Parameter	Estimate	E.S.E.	t-value	p-value
a	-16.241	14.537	-1.117	0.267
b	0.720	0.441	1.632	0.106
c	7.932	8.916	0.890	0.376
d	1.579	1.359	1.162	0.248

Table 9 Computer output obtained for Exercise 3.4

3.5

(i) In neither case does a quadratic term appear necessary. A linear relationship between volume and girth appears reasonable. A plot of volume against height seems to suggest non-homogeneity of variance rather than a quadratic relationship. We start by considering the full model including quadratic terms in girth and height. The t-statistics in Table 10 give the following p-values:

Variable	p-value (full model)	p-value (reduced model)
Girth	0.068	0.036
Height	0.947	0.000
Girth^2	0.000	0.000
Height^2	0.887	–

Table 10 Computer output obtained for Exercise 3.5

Next, we consider the model including only a linear term in h. Here all the t-statistics give significant p-values, see the third column of Table 10, suggesting $h + g + g^2$ is a useful candidate model.

(ii) The model is $v = hg^2$, no intercept, by analogy with the formula for the volume of a cylinder. Results obtained for a variety of models are shown in Table 11. We see that not only is this model more interpretable and more parsimonious, it also produces fitted values in better agreement with the experimental data and leading to lower residual sum of squares.

Model	Residual SS
$I + g$	2.749
$I + h$	27.292
$I + g + h + g^2 + h^2$	0.974
$I + g + h + g^2$	0.975
hg^2	0.948

Table 11   Residual sum of squares for a variety of competing models ('I' denotes intercept).

3.6

(i) $a^T x = \sum_i a_i x_i$, so $\partial(a^T x)/\partial x_r = a_r$. The first part follows on assembling these scalars as a vector, and the second on assembling these vectors as a matrix.

(ii) $x^T A x = \sum_{ij} a_{ij} x_i x_j = \sum_i a_{ii} x_i^2 + \sum_{i \neq j} a_{ij} x_i x_j$. So

$$\partial(x^T A x)/\partial x_r = 2 a_{rr} x_r + \sum_{j \neq r} a_{rj} x_j + \sum_{i \neq r} a_{ir} x_i,$$

as $x_i x_j$ contributes twice, first from $i = r$ and second from $j = r$. Split the first term on the right into two, and use each to complete one of the incomplete sums. Then the right is

$$\sum_j a_{rj} x_j + \sum_i a_{ir} x_i = \sum_i x_i (a_{ir} + a_{ri}) = \left(x^T \left(A + A^T \right) \right)_r,$$

and the result follows on assembling these into a vector.

3.7

$$
\begin{aligned}
SS : &= (y - A\beta)^T (y - A\beta) \\
&= y^T y - y^T A\beta - \beta^T A^T y + \beta^T A^T A\beta \\
&= y^T y - 2 y^T A\beta + \beta^T A^T A\beta,
\end{aligned}
$$

since $y^T A\beta$ is a scalar. Differentiating with respect to β gives

$$\frac{\partial SS}{\partial \beta} = -2 y^T A + 2 \beta^T A^T A,$$

which when set equal to zero and taking the matrix transpose gives

$$(A^T A)\beta = A^T y.$$

3.8

$\|a_1\|^2 = 9$, so normalising gives $q_1 = (-2/3, -1/3, -2/3, 0)^T$ and $a_2 q_1 = -10/3$. This gives

$$w_2 = a_2 - (a_2^T q_1)q_1 = \left(-\frac{2}{9}, \frac{8}{9}, -\frac{2}{9}, 1\right)^T.$$

$\|w_2\|^2 = 153/81$, so appropriate normalisation gives

$$q_2 = \frac{1}{\sqrt{153}}(-2, 8, -2, 9)^T.$$

We have that $(a_3^T q_1) = 8/3$, $(a_3^T q_2) = -19/\sqrt{153}$, $a_3 - (a_3^T q_1)q_1 = a_3 - (8/3)q_1 = (-2/9, -10/9, 7/9, -1)^T$. Similarly, we find that

$$-(a_3^T q_2)q_2 = \frac{1}{153}(-38, 152, -38, 171)^T.$$

Combining gives

$$w_3 = \frac{1}{153}(-55, -18, 81, 18)^T,$$

and on normalisation,

$$q_3 = \frac{153}{\sqrt{10234}}(-55, -18, 81, 18)^T.$$

3.9

We have that

$$X = \begin{pmatrix} 1 & \cdots & 1 \\ x_1 & \cdots & x_n \end{pmatrix}^T = \begin{pmatrix} \mathbf{1} \\ x \end{pmatrix}^T.$$

$q_1 = \mathbf{1}/\sqrt{n}$ where $\mathbf{1}$ denotes a column vector of 1s. $w_2 = a_2 - (a_2^T q_1)q_1 = x - (x^T q_1)q_1 = x - \bar{x}$. This gives

$$q_2 = \frac{x - \bar{x}}{\sqrt{\sum(x_i - \bar{x})^2}}.$$

We have that $a_1 = \sqrt{n}\, q_1$ and $a_2 = \sqrt{n}\, \bar{x} q_1 + \sqrt{\sum(x_i - \bar{x})^2}\, q_2$, which allows us to read off

$$Q = \begin{pmatrix} \frac{1}{\sqrt{n}} & \frac{x - \bar{x}}{\sqrt{\sum(x_i - \bar{x})^2}} \end{pmatrix}, R = \begin{pmatrix} \sqrt{n} & \sqrt{n}\bar{x} \\ 0 & \sqrt{\sum(x_i - \bar{x})^2} \end{pmatrix},$$

$$R^{-1} = \frac{1}{\sqrt{n\sum(x_i - \bar{x})^2}} \begin{pmatrix} \sqrt{\sum(x_i - \bar{x})^2} & -\sqrt{n}\bar{x} \\ 0 & \sqrt{n} \end{pmatrix}.$$

Performing the matrix multiplications gives

$$Q^T y = \begin{pmatrix} \sqrt{n}\,\overline{y} \\ \frac{\sum x_i y_i - n\overline{x}\,\overline{y}}{\sqrt{\sum(x_i - \overline{x})^2}} \end{pmatrix},$$

$$R^{-1}Q^T y = \begin{pmatrix} \overline{y} - \frac{\overline{x}\sum x_i y_i - n\overline{x}\,\overline{y}}{\sqrt{\sum(x_i - \overline{x})^2}} \\ \frac{\sum x_i y_i - n\overline{x}\,\overline{y}}{\sum(x_i - \overline{x})^2} \end{pmatrix}.$$

3.10

(i) This follows from 1. and 2. in Exercise 1.11 since conditionally on u x_i/Δ_{ii} is $N(0, u)$.

(ii) $100(1 - \alpha)\%$ confidence intervals are produced by multiplying estimated standard errors by $t_{n-p}(1 - \alpha/2)$. For the simple linear regression model $p = 2$. We have that

$$\hat{\sigma} = \sqrt{\frac{1}{n-2}\left(S_{yy} - \frac{S_{xy}^2}{S_{xx}}\right)}.$$

$$X^T X = \begin{pmatrix} 1 & x_1 \\ \vdots & \vdots \\ 1 & x_n \end{pmatrix}^T \begin{pmatrix} 1 & x_1 \\ \vdots & \vdots \\ 1 & x_n \end{pmatrix} = \begin{pmatrix} n & n\overline{x} \\ n\overline{x} & \sum x_i^2 \end{pmatrix},$$

$$(X^T X)^{-1} = \frac{1}{nS_{xx}} \begin{pmatrix} \sum x_i^2 & -n\overline{x} \\ -n\overline{x} & n \end{pmatrix}.$$

This gives,

$$\text{e.s.e}(\hat{a}) = \hat{\sigma}\sqrt{\frac{\sum x_i^2}{nS_{xx}}}, \quad \text{e.s.e}(\hat{b}) = \frac{\hat{\sigma}}{\sqrt{S_{xx}}},$$

where e.s.e represents estimated standard error.

(iii) For the bivariate regression model $p = 3$. We have that

$$\hat{\sigma} = \sqrt{\frac{1}{n-3}\left(S_{yy} - aS_{uy} - bS_{vy}\right)}.$$

$$X^T X = \begin{pmatrix} 1 & 1 & 1 \\ \vdots & \vdots & \vdots \\ 1 & u_n & v_n \end{pmatrix}^T \begin{pmatrix} 1 & 1 & 1 \\ \vdots & \vdots & \vdots \\ 1 & u_n & v_n \end{pmatrix} = \begin{pmatrix} n & n\overline{u} & n\overline{v} \\ n\overline{u} & \sum u_i^2 & \sum u_i v_i \\ n\overline{v} & \sum u_i v_i & \sum v_i^2 \end{pmatrix}.$$

The determinant of this matrix can be calculated as

$$|X^T X| = n\left[\sum u_i^2 \sum v_i^2 - \left(\sum u_i v_i\right)\right] - n\overline{u}\left[n\overline{u}\sum v_i^2 - n\overline{v}\sum u_i v_i\right]$$
$$+ n\overline{v}\left[n\overline{u}\sum u_i v_i - n\overline{v}\sum u_i^2\right],$$

which can be simplified slightly to give

$$|X^T X| = n \sum u_i^2 S_{vv} + n \sum u_i v_i \left[2n\overline{uv} - \sum u_i v_i \right] - n^2 \overline{u}^2 \sum v_i^2.$$

In order to calculate estimated standard errors we need the diagonal elements of $(X^T X)^{-1}$. These can be calculated directly as co-factor/determinant without inverting the whole matrix. These calculations then directly give

$$\text{e.s.e}(\hat{c}) = \hat{\sigma} \sqrt{\frac{\sum u_i^2 \sum v_i^2 - (\sum u_i v_i)^2}{n \sum u_i^2 S_{vv} + n \sum u_i v_i \left[2n\overline{uv} - \sum u_i v_i \right] - n^2 \overline{u}^2 \sum v_i^2}},$$

$$\text{e.s.e}(\hat{a}) = \hat{\sigma} \sqrt{\frac{S_{vv}}{\sum u_i^2 S_{vv} + \sum u_i v_i \left[2n\overline{uv} - \sum u_i v_i \right] - n\overline{u}^2 \sum v_i^2}},$$

$$\text{e.s.e}(\hat{b}) = \hat{\sigma} \sqrt{\frac{S_{uu}}{\sum u_i^2 S_{vv} + \sum u_i v_i \left[2n\overline{uv} - \sum u_i v_i \right] - n\overline{u}^2 \sum v_i^2}}.$$

3.11
$E(Y(X_0^T)) = X_0^T \beta$ and $\text{var}(Y(X_0^T)) = \sigma^2$. The appropriate point estimate remains $X_0^T \hat{\beta}$, but the variance associated with this estimate is $\text{var}(X_0^T \hat{\beta}) + \sigma^2 = \sigma^2(1 + X_0^T (X^T X)^{-1} X_0)$. The appropriate confidence interval should thus be constructed as

$$X_0^T \hat{\beta} \pm t_{n-p}(1 - \alpha/2) \hat{\sigma} \sqrt{\left(1 + X_0^T (X^T X)^{-1} X_0 \right)}.$$

3.12

$$(AB)_{ii} = \sum_j A_{ij} B_{ji},$$

$$\text{trace}(AB) = \sum_{i,j} A_{ij} B_{ji} = \sum_{i,j} B_{ji} A_{ij} = \sum_{i,j} B_{ij} A_{ji} = \text{trace}(BA).$$

Chapter 4

4.1
Linear model $Y = a + bX$. The t-test gives a p-value of 0.000, indicating that the X term is needed in the model. The R^2 is a reasonably high 0.65.
Quadratic model $Y = a + bX + cX^2$. The t-test gives a p-value of 0.394, suggesting that the X^2 term is not needed in the model.
Proceeding in this manner, the t-test also suggests a cubic term is not needed in the model. In conclusion, a simple linear regression model seems adequate. The parameter estimates with estimated standard errors are $a = 7.886$ (0.787) and $b = -0.280$ (0.058). The percentage of divorces caused by adultery appears to decrease by 0.28% per year of marriage.

4.2

(i) Result follows using $y = (y_1, y_2, \ldots, y_k)^T$.

(ii) $\hat{\beta} = \begin{pmatrix} \bar{y}_1 & \bar{y}_2 - \bar{y}_1 \end{pmatrix}^T$. Fitted values are

$$\begin{pmatrix} 1 \\ 0 \end{pmatrix}^T \begin{pmatrix} \bar{y}_1 \\ \bar{y}_2 - \bar{y}_1 \end{pmatrix} = \bar{y}_1; \quad \begin{pmatrix} 1 \\ 1 \end{pmatrix}^T \begin{pmatrix} \bar{y}_1 \\ \bar{y}_2 - \bar{y}_1 \end{pmatrix} = \bar{y}_2.$$

(iii) Consider $A^T A M_{ij}$. W.l.o.g. assume $i \leq j$ ($A^T A M$ is symmetric).

$$A^T A M_{ij} = \begin{pmatrix} n_i \\ 0 \\ \cdots \\ 0 \\ n_i \\ 0 \\ \cdots \\ 0 \end{pmatrix}^T \begin{pmatrix} -\frac{1}{n_1} \\ \cdot \\ \cdot \\ \cdot \\ \frac{1}{n_1} \\ \cdot \\ \cdot \\ \cdot \end{pmatrix} = 0,$$

$$A^T A M_{ii} = \begin{pmatrix} n_i \\ 0 \\ \cdots \\ 0 \\ n_i \\ 0 \\ \cdots \\ 0 \end{pmatrix}^T \begin{pmatrix} -\frac{1}{n_1} \\ \cdot \\ \cdot \\ \frac{1}{n_1} + \frac{1}{n_i} \\ \cdot \\ \cdot \end{pmatrix} = 1.$$

Hence $A^T A M = I$. $\hat{\beta} = \begin{pmatrix} \bar{y}_1, & \bar{y}_2 - \bar{y}_1, & \ldots, & \bar{y}_k - \bar{y}_1 \end{pmatrix}^T$. β_1 gives the mean of treatment group 1, which acts as a baseline. β_j is an offset and gives the difference between group j and group 1, so that the mean of group j is $\beta_j + \beta_1$.

4.3

Under a regression formulation (and corner–point constraints) we have that

$$
A = \begin{pmatrix}
1 & 0 & 0 & 0 & 0 & 0 \\
1 & 0 & 0 & 0 & 0 & 0 \\
1 & 0 & 0 & 0 & 0 & 0 \\
1 & 0 & 0 & 0 & 0 & 0 \\
1 & 1 & 0 & 0 & 0 & 0 \\
1 & 1 & 0 & 0 & 0 & 0 \\
1 & 1 & 0 & 0 & 0 & 0 \\
1 & 1 & 0 & 0 & 0 & 0 \\
1 & 0 & 1 & 0 & 0 & 0 \\
1 & 0 & 1 & 0 & 0 & 0 \\
1 & 0 & 1 & 0 & 0 & 0 \\
1 & 0 & 1 & 0 & 0 & 0 \\
1 & 0 & 0 & 1 & 0 & 0 \\
1 & 0 & 0 & 1 & 0 & 0 \\
1 & 0 & 0 & 1 & 0 & 0 \\
1 & 0 & 0 & 1 & 0 & 0 \\
1 & 0 & 0 & 0 & 1 & 0 \\
1 & 0 & 0 & 0 & 1 & 0 \\
1 & 0 & 0 & 0 & 1 & 0 \\
1 & 0 & 0 & 0 & 1 & 0 \\
1 & 0 & 0 & 0 & 0 & 1 \\
1 & 0 & 0 & 0 & 0 & 1 \\
1 & 0 & 0 & 0 & 0 & 1 \\
1 & 0 & 0 & 0 & 0 & 1
\end{pmatrix}, \quad
A^T A = \begin{pmatrix}
24 & 4 & 4 & 4 & 4 & 4 \\
4 & 4 & 0 & 0 & 0 & 0 \\
4 & 0 & 4 & 0 & 0 & 0 \\
4 & 0 & 0 & 4 & 0 & 0 \\
4 & 0 & 0 & 0 & 4 & 0 \\
4 & 0 & 0 & 0 & 0 & 4
\end{pmatrix},
$$

$$
(A^T A)^{-1} = \begin{pmatrix}
0.25 & -0.25 & -0.25 & -0.25 & -0.25 & -0.25 \\
-0.25 & 0.5 & 0.25 & 0.25 & 0.25 & 0.25 \\
-0.25 & 0.25 & 0.5 & 0.25 & 0.25 & 0.25 \\
-0.25 & 0.25 & 0.25 & 0.5 & 0.25 & 0.25 \\
-0.25 & 0.25 & 0.25 & 0.25 & 0.5 & 0.25 \\
-0.25 & 0.25 & 0.25 & 0.25 & 0.25 & 0.5
\end{pmatrix},
$$

$$
A^T y = \begin{pmatrix}
266.5 \\
41 \\
46.5 \\
47 \\
42 \\
48
\end{pmatrix}, \quad
\hat{\beta} = \begin{pmatrix}
10.5 \\
-0.25 \\
1.125 \\
1.25 \\
0 \\
1.5
\end{pmatrix}.
$$

4.4

$$A = \begin{pmatrix}
1 & 0 & 0 & 0 & 0 & 0 & 0 & 0 & 0 \\
1 & 0 & 0 & 0 & 0 & 0 & 1 & 0 & 0 \\
1 & 0 & 0 & 0 & 0 & 0 & 0 & 1 & 0 \\
1 & 0 & 0 & 0 & 0 & 0 & 0 & 0 & 1 \\
1 & 1 & 0 & 0 & 0 & 0 & 0 & 0 & 0 \\
1 & 1 & 0 & 0 & 0 & 0 & 1 & 0 & 0 \\
1 & 1 & 0 & 0 & 0 & 0 & 0 & 1 & 0 \\
1 & 1 & 0 & 0 & 0 & 0 & 0 & 0 & 1 \\
1 & 0 & 1 & 0 & 0 & 0 & 0 & 0 & 0 \\
1 & 0 & 1 & 0 & 0 & 0 & 1 & 0 & 0 \\
1 & 0 & 1 & 0 & 0 & 0 & 0 & 1 & 0 \\
1 & 0 & 1 & 0 & 0 & 0 & 0 & 0 & 1 \\
1 & 0 & 0 & 1 & 0 & 0 & 0 & 0 & 0 \\
1 & 0 & 0 & 1 & 0 & 0 & 1 & 0 & 0 \\
1 & 0 & 0 & 1 & 0 & 0 & 0 & 1 & 0 \\
1 & 0 & 0 & 1 & 0 & 0 & 0 & 0 & 1 \\
1 & 0 & 0 & 0 & 1 & 0 & 0 & 0 & 0 \\
1 & 0 & 0 & 0 & 1 & 0 & 1 & 0 & 0 \\
1 & 0 & 0 & 0 & 1 & 0 & 0 & 1 & 0 \\
1 & 0 & 0 & 0 & 1 & 0 & 0 & 0 & 1 \\
1 & 0 & 0 & 0 & 0 & 1 & 0 & 0 & 0 \\
1 & 0 & 0 & 0 & 0 & 1 & 1 & 0 & 0 \\
1 & 0 & 0 & 0 & 0 & 1 & 0 & 1 & 0 \\
1 & 0 & 0 & 0 & 0 & 1 & 0 & 0 & 1
\end{pmatrix},$$

and we calculate

$$A^T A = \begin{pmatrix}
24 & 4 & 4 & 4 & 4 & 4 & 6 & 6 & 6 \\
4 & 4 & 0 & 0 & 0 & 0 & 1 & 1 & 1 \\
4 & 0 & 4 & 0 & 0 & 0 & 1 & 1 & 1 \\
4 & 0 & 0 & 4 & 0 & 0 & 1 & 1 & 1 \\
4 & 0 & 0 & 0 & 4 & 0 & 1 & 1 & 1 \\
4 & 0 & 0 & 0 & 0 & 4 & 1 & 1 & 1 \\
6 & 1 & 1 & 1 & 1 & 1 & 6 & 0 & 0 \\
6 & 1 & 1 & 1 & 1 & 1 & 0 & 6 & 0 \\
6 & 1 & 1 & 1 & 1 & 1 & 0 & 0 & 6
\end{pmatrix},$$

with $(A^T A)^{-1}$ given by

$$
\begin{pmatrix}
3/8 & -1/4 & -1/4 & -1/4 & -1/4 & -1/4 & -1/6 & -1/6 & -1/6 \\
-1/4 & 1/2 & 1/4 & 1/4 & 1/4 & 1/4 & 0 & 0 & 0 \\
-1/4 & 1/4 & 1/2 & 1/4 & 1/4 & 1/4 & 0 & 0 & 0 \\
-1/4 & 1/4 & 1/4 & 1/2 & 1/4 & 1/4 & 0 & 0 & 0 \\
-1/4 & 1/4 & 1/4 & 1/4 & 1/2 & 1/4 & 0 & 0 & 0 \\
-1/4 & 1/4 & 1/4 & 1/4 & 1/4 & 1/2 & 0 & 0 & 0 \\
-1/6 & 0 & 0 & 0 & 0 & 0 & 1/3 & 1/6 & 1/6 \\
-1/6 & 0 & 0 & 0 & 0 & 0 & 1/6 & 1/3 & 1/6 \\
-1/6 & 0 & 0 & 0 & 0 & 0 & 1/6 & 1/6 & 1/3
\end{pmatrix},
$$

$$
A^T y = \begin{pmatrix}
266.5 \\
41 \\
46.5 \\
47 \\
42 \\
48 \\
71.5 \\
67.5 \\
47.5
\end{pmatrix}, \quad
\hat{\beta} = \begin{pmatrix}
12.729 \\
-0.25 \\
1.125 \\
1.25 \\
0 \\
1.5 \\
-1.417 \\
-2.083 \\
-5.417
\end{pmatrix}.
$$

4.5

(i) $N_3(0, \Sigma)$, $\Sigma = \begin{pmatrix} \sigma_0^2 + \sigma^2 & \sigma_0^2 + \sigma^2 & \sigma_0^2 + \sigma^2 \\ \sigma_0^2 + \sigma^2 & \sigma_0^2 + 2\sigma^2 & \sigma_0^2 + 2\sigma^2 \\ \sigma_0^2 + \sigma^2 & \sigma_0^2 + 2\sigma^2 & \sigma_0^2 + 3\sigma^2 \end{pmatrix}$

(ii) $N_n(0, \Sigma)$, $\Sigma_{jj} = \sigma_0^2 + j\sigma^2$, $\Sigma_{ij} = \sigma_0^2 + \min(i, j)\sigma^2$.

4.6

Let

$$
X = \begin{pmatrix} 1 & 1 & 1 \\ 1 & -1 & -1 \end{pmatrix} \begin{pmatrix} Y_1 \\ Y_2 \\ Y_3 \end{pmatrix}.
$$

By the linear transformation property, X is multivariate normal with mean vector

$$
\begin{pmatrix} 1 & 1 & 1 \\ 1 & -1 & -1 \end{pmatrix} \begin{pmatrix} \mu_1 \\ \mu_2 \\ \mu_3 \end{pmatrix} = \begin{pmatrix} \mu_1 + \mu_2 + \mu_3 \\ \mu_1 - \mu_2 - \mu_3 \end{pmatrix},
$$

and covariance matrix

$$\begin{pmatrix} 1 & 1 \\ 1 & -1 \\ 1 & -1 \end{pmatrix}^T \begin{pmatrix} 1 & a & 0 \\ a & 1 & b \\ 0 & b & 1 \end{pmatrix} \begin{pmatrix} 1 & 1 \\ 1 & -1 \\ 1 & -1 \end{pmatrix} = \begin{pmatrix} 3 + 2a + 2b & -1 - 2b \\ -1 - 2b & 3 + 2b - 2a \end{pmatrix}.$$

The components of X are independent iff the covariance matrix is diagonal, which occurs iff $b = -1/2$.

4.7

(i)

$$E(Y^n) = \frac{b^{n+1} + (-1)^n a^{n+1}}{(n+1)(a+b)}.$$

(ii) $E(X|Y = y) = y^2$. $E(Y|X = x) = ((b-a)\sqrt{x})/(a+b)$.
(iii) The problem is to find α and β to minimise $E(Y^2 - \alpha - \beta Y)^2$ and $E(Y - \alpha - \beta Y^2)^2$.

$$E(Y^2 - \alpha - \beta Y)^2 = E(Y^4 - 2\alpha Y^2 - 2\beta Y^3 + \alpha^2 + \beta^2 Y^2 + 2\alpha\beta Y),$$

which gives

$$\frac{b^5 + a^5}{5(b+a)} + \frac{(\beta^2 - 2\alpha)b^3 + a^3}{3(b+a)} - \frac{2\beta(b^4 - a^4)}{4(b+a)} + \alpha^2 + \alpha\beta(b-a).$$

Differentiating and equating to zero gives

$$\begin{pmatrix} 2 & b-a \\ b-a & \frac{2(b^3+a^3)}{3(b+a)} \end{pmatrix} \begin{pmatrix} \alpha \\ \beta \end{pmatrix} = \begin{pmatrix} \frac{2(b^3+a^3)}{3(b+a)} \\ \frac{(b^4-a^4)}{2(b+a)} \end{pmatrix},$$

with solution

$$\alpha = \frac{4(b^2 - ab + a^2)(b^3 + a^3)}{3(a+b)^3} - \frac{3(b-a)(b^4 - a^4)}{2(a+b)^3},$$

$$\beta = \frac{3(b^4 - a^4)}{(a+b)^3} - \frac{2(b-a)(b^3 + a^3)}{(a+b)^3}.$$

In the second case similar working gives

$$\begin{pmatrix} 2 & \frac{2(b^3+a^3)}{3(b+a)} \\ \frac{2(b^3+a^3)}{3(b+a)} & \frac{2(b^5+a^5)}{5(b+a)} \end{pmatrix} \begin{pmatrix} \alpha \\ \beta \end{pmatrix} = \begin{pmatrix} b-a \\ \frac{(b^4-a^4)}{2(b+a)} \end{pmatrix},$$

with solution

$$\alpha = \frac{3(b^4 - a^4)}{4(a + b)(b^2 - ab - 3 + a^2)} - \frac{3(b - a)}{2(b^2 - ab - 3 + a^2)},$$

$$\beta = \frac{3(b - a)}{2(b^2 - ab - 3 + a^2)} - \frac{9(b^4 - a^4)}{4(b + a)(b^2 - ab + a^2)(b^2 - ab - 3 + a^2)}.$$

4.8

The log-likelihood function is

$$l = \frac{nr}{2} \log(2\pi) + \frac{n}{2} \log|\Lambda| - \frac{1}{2} \sum_{i=1}^{n} (x_i - \mu_0)^T \Lambda (x_i - \mu_0),$$

where r is the dimension. We have that

$$\frac{\partial l}{\partial \Lambda} = \frac{n\Sigma}{2} - \frac{1}{2} \sum_{i=1}^{n} (x_i - \mu_0)^T (x_i - \mu_0).$$

Equating to zero gives

$$\hat{\Sigma} = \frac{\sum_{i=1}^{n} (x_i - \mu_0)^T (x_i - \mu_0)}{n}.$$

Continuing, we see that

$$\begin{aligned}
\hat{\Sigma} &= \sum_{i=1}^{n} \frac{(x_i - \bar{x} - \mu_0 + \bar{x})^T (x_i - \bar{x} - \mu_0 + \bar{x})}{n}, \\
&= \sum_{i=1}^{n} \frac{(x_i - \bar{x})^T (x_i - \bar{x}) + (\bar{x} - \mu_0)^T (x_i - \bar{x} - \mu_0 + \bar{x})}{n}, \\
&= S + \frac{n(\bar{x} - \mu_0)}{n},
\end{aligned}$$

since $\sum_{i=1}^{n} (x_i - \bar{x}) = 0$.

4.9

(i)

$$f_X(x) \propto \exp\left\{ -\frac{1}{2} (x - \mu)^T \Sigma^{-1} (x - \mu) \right\},$$

from which it follows that $A = -1/2\Sigma^{-1}$ and $b = -(1/2)(-2)\Sigma^{-1}\mu$.

(ii) We have that

$$f_{X|Y}(x|y) = \frac{f_{X,Y}(x, y)}{f_Y(y)},$$

giving $K = 1/f_Y(y)$.

4.10

We calculate the matrix product

$$\left(\begin{array}{cc} A & B \\ C & D \end{array}\right)\left(\begin{array}{cc} M & -MBD^{-1} \\ -D^{-1}CM & D^{-1}+D^{-1}CMBD^{-1} \end{array}\right),$$

as

$$\left(\begin{array}{cc} AM-BD^{-1}CM & [I-AM+BD^{-1}CM]BD^{-1} \\ CM-DD^{-1}CM & -CMBD^{-1}+I+CMBD^{-1} \end{array}\right)=\left(\begin{array}{cc} I & 0 \\ 0 & I \end{array}\right).$$

Similarly the matrix product

$$\left(\begin{array}{cc} M & -MBD^{-1} \\ -D^{-1}CM & D^{-1}+D^{-1}CMBD^{-1} \end{array}\right)\left(\begin{array}{cc} A & B \\ C & D \end{array}\right)$$

is

$$\left(\begin{array}{cc} M[A-BD^{-1}C] & MB-MB \\ D^{-1}C[I-I] & -D^{-1}CMB+I+D^{-1}CMB \end{array}\right)=\left(\begin{array}{cc} I & 0 \\ 0 & I \end{array}\right).$$

4.11

Define $\Lambda = \Sigma^{-1}$. From the formula for inverses of partitioned matrices we see that Λ can be partitioned as

$$\begin{aligned} \Lambda_{11} &= \left(\Sigma_{AA}-\Sigma_{AB}\Sigma_{BB}^{-1}\Sigma_{BA}\right)^{-1}, \\ \Lambda_{12} &= -\left(\Sigma_{AA}-\Sigma_{AB}\Sigma_{BB}^{-1}\Sigma_{BA}\right)^{-1}\Sigma_{AB}\Sigma_{BB}^{-1}, \\ \Lambda_{21} &= -\Sigma_{BB}^{-1}\Sigma_{BA}\left(\Sigma_{AA}-\Sigma_{AB}\Sigma_{BB}^{-1}\Sigma_{BA}\right)^{-1}, \\ \Lambda_{22} &= \Sigma_{BB}^{-1}+\Sigma_{BB}^{-1}\Sigma_{BA}\left(\Sigma_{AA}-\Sigma_{AB}\Sigma_{BB}^{-1}\Sigma_{BA}\right)^{-1}\Sigma_{AB}\Sigma_{BB}^{-1}. \end{aligned}$$

$$f_{x_A|x_B} \propto \exp\left(-\frac{1}{2}x_A^T\Lambda_{AA}x_A + x_A^T\Lambda_{AA}\mu_A - x_A^T\Lambda_{AB}(x_B-\mu_B)\right).$$

It follows that

$$\Sigma_{A|B}=\Lambda_{AA}^{-1}=\Sigma_{AA}-\Sigma_{AB}\Sigma_{BB}^{-1}\Sigma_{BA}.$$

Continuing, we see that

$$\Sigma_{A|B}^{-1}\mu_{A|B}=\Lambda_{AA}\mu_A-\Lambda_{AB}(x_B-\mu_B),$$

giving

$$\mu_{A|B}=\Lambda_{AA}^{-1}\left(\Lambda_{AA}\mu_A-\Lambda_{AB}(x_B-\mu_B)\right)=\mu_A+\Sigma_{AB}\Sigma_{BB}^{-1}(x_B-\mu_B).$$

Source	df	SS	Mean Square	F	p
Age	1	17.269	17.269	4.165	0.069
Club membership	1	74.805	74.805	17.674	0.002
Age:Club membership (Different slopes)	1	1.355	1.355	0.320	0.584
Residual	10	42.324	4.232		
Total	13				

Table 12 Exercise 5.1. ANOVA table for model with different intercepts and different slopes

Chapter 5

5.1
The completed ANOVA table is shown in Table 12.

5.2
We have that $\mathrm{var}(\hat{\lambda}) = \sigma^2(V^T V)^{-1}$, which gives $\mathrm{var}(\hat{\gamma}_A) = \sigma^2(Z^T R Z)^{-1}$. Since $\mathrm{cov}(\hat{\alpha}, \gamma) = \mathrm{cov}(\hat{\beta}, \hat{\gamma}) + (X^T X)^{-1}(X^T Z)\mathrm{var}(\hat{\gamma}) = 0$, this gives $\mathrm{cov}(\hat{\beta}, \hat{\gamma}) = -\sigma^2(X^T X)^{-1}(X^T Z)(Z^T R Z)$, with $\mathrm{cov}(\hat{\gamma}, \hat{\beta}) = \mathrm{cov}(\hat{\beta}, \hat{\gamma})^T$. Finally, since $\hat{\beta} = \hat{\alpha} - (X^T X)^{-1} X^T Z \gamma$, with both $\hat{\alpha}$ and $\hat{\beta}$ independent, we can calculate $\mathrm{var}(\hat{\beta})$ as

$$
\begin{aligned}
\mathrm{var}(\hat{\beta}) &= \mathrm{var}(\hat{\alpha}) + (X^T X)^{-1} X^T Z \mathrm{var}(\hat{\gamma}) Z^T X (X^T X)^{-1}, \\
&= (X^T X)^{-1} + (X^T X)^{-1} X^T Z (Z^T R Z)^{-1} Z^T X (X^T X)^{-1}.
\end{aligned}
$$

5.3
(i) $\hat{\alpha} = \overline{Y}$.
(ii) We have that $RY = (Y_1 - \overline{Y}, \ldots, Y_n - \overline{Y})^T$, $X = (1, \ldots, 1)^T$, $x_p = (x_1, \ldots, x_n)^T$. We have that

$$
\hat{\beta}_A = \frac{x_{(p)}^T R Y}{x_{(p)}^T R x_{(p)}} = \frac{\begin{pmatrix} x_1 \\ \vdots \\ x_n \end{pmatrix}^T \begin{pmatrix} Y_1 - \overline{Y} \\ \vdots \\ Y_n - \overline{Y} \end{pmatrix}}{\begin{pmatrix} x_1 \\ \vdots \\ x_n \end{pmatrix}^T \begin{pmatrix} x_1 - \overline{x} \\ \vdots \\ x_n - \overline{x} \end{pmatrix}} = \frac{S_{x_{(p)} Y}}{S_{x_{(p)} x_{(p)}}},
$$

$$
\hat{\alpha}_A = \overline{Y} - (X^T X)^{-1} X^T x_{(p)} \hat{\beta}_A,
$$

$$= \overline{Y} - \frac{1}{n} \begin{pmatrix} 1 \\ \vdots \\ 1 \end{pmatrix}^{T} \begin{pmatrix} x_1 \\ \vdots \\ x_n \end{pmatrix} \hat{\beta}_A,$$

$$= \overline{Y} - \frac{1}{n} n \overline{x}_{(p)} \hat{\beta}_A = \overline{Y} - \overline{x}_{(p)} \hat{\beta}_A.$$

5.5
Fitting the model with different slopes leads to a residual sum of squares of 13902 on 19 df The model with different slopes leads to a residual sum of squares of 14206.4 on 20 df We obtain that $F = 0.416$ and a p-value of 0.527 when comparing against $F_{1,19}$. Fitting the model with just a single slope gives a residual sum of squares of 17438.9 on 21 df We obtain that $F = 4.551$ and a p-value of 0.045 when comparing against $F_{1,20}$. Thus, two different intercepts are required but not two different slopes.

5.6
The plot shows definite signs of curvature. The assumption of a linear relationship on the log-scale seems altogether more reasonable, although there is still perhaps an element of curvature that remains. On the log scale, fitting the model with different slopes leads to a residual sum of squares of 0.365 on 19 df The model with different slopes leads to a residual sum of squares of 0.379 on 20 df We obtain $F = 0.746$ and a p-value of 0.399 when comparing against $F_{1,19}$. Fitting the model with just a single slope gives a residual sum of squares of 0.546 on 21 df We obtain $F = 8.801$ and a p-value of 0.008 when comparing against $F_{1,20}$. The fitted equation reads

$$\text{rate} = 220.0027 \times 0.843 I(\text{untreated}) \text{concentration}^{0.287},$$

where $I(\cdot)$ is an indicator representing lack of treatment. Thus there is an approximate power-law relationship between rate and concentration and treatment does increase the rate of reaction.

5.7
Fitting the model with different slopes gives a residual sum of squares of 4.630 on 41 df. Further, t-tests for all the different intercept terms are significant, suggesting all these terms really are needed in the model. Fitting the model with different slopes gives a residual sum of squares of 1.369 on 35 df We obtain $F = 13.895$ and a p-value of 0.000 when compared to $F_{6,41}$. Using corner-point constraints, parameter estimates for Asia and Africa are both positive, and using the t-test are borderline significant and significant respectively. This suggests the rate of increase in phone usage is roughly the same across the world, but rapidly increasing in both Asia and Africa.

5.8

(i)

$$X = \begin{pmatrix} 1 & x_1 & x_1^2 & 0 & 0 & 0 \\ \vdots & \vdots & \vdots & \vdots & \vdots & \vdots \\ 1 & x_k & x_k^2 & 0 & 0 & 0 \\ 1 & x_{k+1} & x_{k+1}^2 & 1 & x_{k+1} & x_{k+1}^2 \\ \vdots & \vdots & \vdots & \vdots & \vdots & \vdots \\ 1 & x_n & x_n^2 & 1 & x_n & x_n^2 \end{pmatrix}.$$

(ii) This is a test of the hypothesis $\beta_2 = \gamma_2 = 0$. Let $SSE_{\beta_2 \neq 0}$ be the residual sum of squares for the full quadratic Analysis of Covariance model, $SSE_{\beta_2 = \gamma_2 = 0}$ be the residual sum of squares for the Analysis of Covariance model with different slopes and different intercepts. The appropriate F-test is

$$\frac{(SSE_{\beta_2 = \gamma_2 = 0} - SSE_{\beta_2 \neq 0})/2}{(SSE_{\beta_2 \neq 0})/(n - 6)} \sim F_{2, n-6}.$$

(iii) Let $SSE_{\beta_2 = 0}$ be the residual sum of squares for the full Analysis of Covariance model without a separate quadratic term. The appropriate F-test is

$$\frac{SSE_{\beta_2 = 0} - SSE_{\beta_2 \neq 0}}{(SSE_{\beta_2 \neq 0})/(n - 6)} \sim F_{1, n-6}.$$

5.9

(i) Set

$$\frac{i - \frac{1}{2}}{n} = \Phi\left(\frac{x_i - \mu}{\sigma}\right), \quad \Phi^{-1}\left(\frac{i - \frac{1}{2}}{n}\right) = \frac{x_i - \mu}{\sigma}.$$

Plot $\Phi^{-1}((i - 1/2)/n)$ against x_i. This should be a straight line with slope $1/\sigma$ and intercept μ/σ.

(ii) Let $Y_i = \Phi^{-1}((i - 1/2)/n)$ and X denote the simulated sample. The plot appears reasonably linear. As a rough guide, t-tests give a p-value of 0.217 that the intercept is zero and a p-value of 0.011 that the slope is 1. So the plot appears reasonably linear, but the slope seems a little far away from the theoretical value of 1 that we might expect.

(iii) $a = (b + 1)L^{-(b+1)}$, from which it follows that $F(x) = (x/L)^{b+1}$. Arrange the x_i into increasing order of size. Set

$$\frac{i - \frac{1}{2}}{n} = \left(\frac{x_i}{L}\right)^{b+1}, \quad \log\left(\frac{i - \frac{1}{2}}{n}\right) = (b + 1)\log x_i - (b + 1)\log L.$$

Plot $\log((i - 1/2)/n)$ against $\log x_i$. The resulting plot should be a straight line with slope $b + 1$ and intercept $-(b + 1) \log L$.

5.10
(i) Write $Y = \beta_0 + \beta_1 X + \gamma_0 Z + \gamma_1 XZ + \epsilon$, where Z is a dummy variable indicating whether or not $X > 4$. The X-matrix is

$$X = \begin{pmatrix} 1 & 1 & 1 & 1 & 1 & 1 & 1 & 1 & 1 \\ 1 & 2 & 3 & 4 & 5 & 6 & 7 & 8 & 9 \\ 0 & 0 & 0 & 0 & 1 & 1 & 1 & 1 & 1 \\ 0 & 0 & 0 & 0 & 5 & 6 & 7 & 8 & 9 \end{pmatrix}^T.$$

Using $\hat{\beta} = (X^T X)^{-1} X^T y$ gives the estimates $(-0.15, 2.05, 3.74, -0.92)^T$. The first fitted straight line is $y = -0.15 + 2.05x$. The second fitted straight line is $y = (-0.15 + 3.75) + (2.05 - 0.92)x = 3.59 + 1.13x$.
(ii) Using trial and error, a simple table of residual sums of squares (Table 13) suggests that the model with a change-point at $x = 4$ is best.

Change-point	4	5	6	7
SSE	0.615	0.686	1.037	1.118

Table 13 Change-point models for Exercise 5.10

(iii) The likelihood function can be written as

$$\frac{1}{\left(\sqrt{2\pi\hat{\sigma}^2}\right)^n} \exp\left\{ -\frac{1}{2\hat{\sigma}^2} \sum_{i=1}^{n} (y_i - X_i^T \hat{\beta})^T (y_i - X_i^T \hat{\beta}) \right\},$$

which we write as

$$\left[\frac{2\pi SSE}{n} \right]^{-\frac{n}{2}} \exp\left\{ \frac{-SSE}{\frac{2SSE}{n}} \right\} = \left(\frac{2\pi e}{n} \right)^{-\frac{n}{2}} SSE^{-\frac{n}{2}}.$$

It follows that the AIC is equivalent to the penalty function

$$-2\left(-\frac{n}{2} \ln(SSE) \right) + 2(p + 1) \propto n\ln(SSE) + 2p.$$

The residual sums of squares for the linear, quadratic and change-point models are 3.509, 1.072 and 0.615, giving 15.298, 6.626 and 3.625 respectively. The change-point model is best.

Chapter 6

6.1

Stepwise regression and forward selection both choose the model

$$\text{Mpg=Weight+Cylinders+Hp.}$$

Alternatively, backward selection chooses the model

$$\text{Mpg=Weight+(1/4)M time+Transmission.}$$

The models are slightly different but both seem reasonable and to offer sensible interpretations.

6.2

The first step is to choose the one-variable model with the highest F statistic given by

$$F = \frac{K - x}{\sqrt{\frac{x}{n-2}}},$$

where K denotes the residual sum-of-squares of the null model and x denotes the residual sum of squares of the candidate model. We have that

$$\frac{dF}{dx} = \sqrt{n - 2} \left(-\frac{1}{2} K x^{-\frac{3}{2}} - \frac{1}{2} x^{-\frac{1}{2}} \right),$$

which is negative and hence a decreasing function of x. Thus we choose the model with lowest residual sum of squares, or equivalently the model with the highest regression sum of squares. In a simple linear regression model we have $SSR = \text{corr}(X, Y)^2$, which gives the result.

6.3

(i)

$$
\begin{aligned}
\#\text{models} &= \#1\text{-term models} + \#2\text{-term models} + \ldots + \#p\text{-term models}, \\
&= \binom{p}{1} + \binom{p}{2} + \cdots + \binom{p}{p}, \\
&= (1 + 1)^n - 1,
\end{aligned}
$$

by the Binomial Theorem.

(ii) 2047.

(iii) Since $2^6 = 64$ and $2^7 = 128$, too large for comfort, this suggests all-subsets regression becomes infeasible in problems with seven or more explanatory variables (in addition to the constant term).

6.4

Set $\partial L/\partial x = \partial L/\partial y = 0$. This gives $y + 2\lambda x = 0$ and $x + 16\lambda y = 0$. This gives two possible cases to consider: $\lambda = -y/(2x)$ and $\lambda = -x/16y$. Either gives the right solution. Here we concentrate upon $\lambda = -y/(2x)$. This gives

$$L = \frac{xy}{2} - \frac{4y^3}{x} + \frac{2y}{x},$$

$$\frac{\partial L}{\partial x} = \frac{y}{2} + \frac{4y^3 - 2y}{x^2},$$

$$\frac{\partial L}{\partial y} = \frac{x}{2} - \frac{12y^2}{x} + \frac{2}{x}.$$

$\partial L/\partial x = 0$ gives $y(x^2/2 + 4y^2 - 2) = 0$ or $y^2 = 1/2 - x^2/8$. $\partial L/\partial y = 0$ then gives $x^2/2 - 12[1/2 - x^2/8] + 2 = 0$ or $x^2 = 2$. The final solution to our problem becomes $\pm(\sqrt{2}, 1/2)$.

6.5

We seek to minimise the functional

$$S := (Y_1 - \alpha)^2 + (Y_2 - \beta)^2 + (Y_3 - \gamma)^2 + \lambda(\pi - \alpha - \beta - \gamma).$$

Differentiating with respect to each of α, β, γ and equating to zero gives

$$-2(Y - 1 - \alpha) = -2(Y_2 - \beta) = -2(Y - 3 - \gamma) = \lambda$$

and sum to obtain

$$\lambda = \frac{2}{3}[\pi - Y_1 - Y_2 - Y_3].$$

The functional S becomes

$$S := (Y_1 - \alpha)^2 + (Y_2 - \beta)^2 + (Y_3 - \gamma)^2 + \frac{2}{3}[\pi - Y_1 - Y_2 - Y_3][\pi - \alpha - \beta - \gamma].$$

The problem reduces to three one-dimensional problems with solution

$$\alpha = \frac{\pi}{3} + \frac{2}{3}Y_1 - \frac{1}{3}Y_2 - \frac{1}{3}Y_3,$$

$$\beta = \frac{\pi}{3} + \frac{2}{3}Y_2 - \frac{1}{3}Y_1 - \frac{1}{3}Y_3,$$

$$\gamma = \frac{\pi}{3} + \frac{2}{3}Y_3 - \frac{1}{3}Y_1 - \frac{1}{3}Y_2.$$

6.6

We seek to minimise the following functional:

$$S := (Y_1 - \alpha)^2 + (Y_2 - \beta)^2 + (Y_3 - \gamma)^2 + (Y_4 - \delta)^2 + \lambda(2\pi - \alpha - \beta - \gamma - \delta).$$

Differentiating with respect to each of α, β, γ, δ in turn and equating to zero gives

$$-2(Y_1 - \alpha) = -2(Y_2 - \beta) = -2(Y_3 - \gamma) = -2(Y_4 - \delta) = \lambda.$$

Summing,

$$4\lambda = -2(Y_1 + Y_2 + Y_3 + Y_4) + 2(\alpha + \beta + \gamma + \delta) = 2(Y_1 + Y_2 + Y_3 + Y_4) + 4\pi,$$

giving

$$\lambda = \pi - \frac{(Y_1 + Y_2 + Y_3 + Y_4)}{2}.$$

The functional S becomes

$$
\begin{aligned}
S : \quad = \quad & (Y_1 - \alpha)^2 + (Y_2 - \beta)^2 + (Y_3 - \gamma)^2 + (Y_4 - \delta)^2 \\
+ \quad & \left(\pi - \frac{(Y_1 + Y_2 + Y_3 + Y_4)}{2} \right) (2\pi - \alpha - \beta - \gamma - \delta).
\end{aligned}
$$

The remaining optimisation reduces to four one-dimensional problems, giving

$$
\begin{aligned}
\alpha &= \pi/2 + (3/4)Y_1 - (1/4)Y_2 - (1/4)Y_3 - (1/4)Y_4, \\
\beta &= \pi/2 + (3/4)Y_2 - (1/4)Y_1 - (1/4)Y_3 - (1/4)Y_4, \\
\gamma &= \pi/2 + (3/4)Y_3 - (1/4)Y_1 - (1/4)Y_2 - (1/4)Y_4, \\
\delta &= \pi/2 + (3/4)Y_4 - (1/4)Y_1 - (1/4)Y_2 - (1/4)Y_3.
\end{aligned}
$$

6.7

We saw in Exercise 4.2 that a regression treatment of the one-way ANOVA model leads to the fitted values $\overline{Y}_i$ for observations from treatment i. The null model has fitted value $\overline{Y}$. The additional sum of squares captured by the treatments model is given by the difference in residual sum of squares:

$$
\begin{aligned}
\sum_{i=1}^{r} \sum_{j=1}^{n_i} \left[(Y_{ij} - \overline{Y})^2 - (Y_{ij} - \overline{Y}_i)^2 \right] &= \sum_{i=1}^{r} -2\overline{Y} n_i \overline{Y}_i + n_i \overline{Y}^2 + 2n_i \overline{Y}_i^2 - n_i \overline{Y}_i^2, \\
&= \sum_{i=1}^{r} n_i (\overline{Y}_i - \overline{Y})^2, \\
&= \sum_{i=1}^{r} n_i \left(\frac{T_i}{n_i} - \frac{T}{n} \right)^2, \\
&= \sum_{i} \frac{T_i^2}{n_i} - \frac{T^2}{n}, \\
&= SST.
\end{aligned}
$$

The residual sum of squares for the larger model is

$$
\begin{aligned}
\sum_{i=1}^{r}\sum_{j=1}^{n_i}(Y_{ij}-\overline{Y}_i)^2 &= \sum_{i=1}^{r}\sum_{j=1}^{n_i}Y_{ij}^2 - \sum_{i=1}^{r}\overline{Y}_i^2, \\
&= \sum_{i,j}Y_{ij}^2 - \sum_{i=1}^{r}\frac{T_i^2}{n}, \\
&= SSE.
\end{aligned}
$$

The F-statistic becomes MST/MSE which is $F_{r-1,n-r}$ under the null hypothesis, as before.

6.8

Fitting a regression model with fertiliser as a factor variable gives a residual sum of squares of 148.188 on 18 df The null model has a residual sum of squares of 159.990 on 23 df Our F-statistic becomes $(1/5)(159.990 - 148.188)/8.233 = 0.287$, giving a p-value of 0.914 as before.

6.9

(i) **Forward selection:**
$(1/4)$M t=v/s+gears+Hp+Weight+Cylinders+Transmission.
(ii) **Backward selection:**
$(1/4)$Mt=Cylinders+Displacement+Weight+v/s+Transmission+
    Carburettors.
(iii) **Stepwise regression:**
$(1/4)$M t=v/s+Hp+Weight+Cylinders+Transmission.

6.10

(i) Fit the model without an intercept term.

$$
\begin{aligned}
y_i &= \beta_0 + \beta_1 x_{1,i} + \ldots + \beta_p x_{p,i} + \epsilon_i, \\
&= \beta_0(x_{1,i} + x_{2,i} + \ldots + x_{p,i}) + \beta_1 x_{1,i} + \ldots + \beta_p x_{p,i} + \epsilon_i, \\
&= (\beta_0 + \beta_1)\,x_{1,i} + (\beta_0 + \beta_2)\,x_{2,i} + \ldots + (\beta_0 + \beta_p)\,x_{p,i} + \epsilon_i.
\end{aligned}
$$

(ii) As (i) we can write $1 = x_1 + x_2 + x_3$. Further $x_1^2 = x_1(1 - x_2 - x_3)$, $x_2^2 = x_2(1 - x_1 - x_3)$, $x_3^2 = x_3(1 - x_1 - x_2)$. The full quadratic model can thus be written

$$
\begin{aligned}
y &= (\beta_0 + \beta_1)x_1 + (\beta_0 + \beta_2)x_2 + (\beta_0 + \beta_3)x_3 + (\beta_{12} - \beta_{11} - \beta_{22})x_1x_2 \\
&+ (\beta_{13} - \beta_{11} - \beta_{33})x_1x_3 + (\beta_{23} - \beta_{22} - \beta_{33})x_2x_3 + \epsilon.
\end{aligned}
$$

For general p the quadratic model becomes

$$
y = \beta_1 x_1 + \beta_2 x_2 + \ldots + \beta_p x_p + \beta_{12}x_1x_2 + \ldots + \beta_{p-1\,p}x_{p-1}x_p + \epsilon.
$$

6.11

(i) The quadratic terms are significant. An Analysis of Variance is shown in Table 14.

Source	df	SS	MS	F	p
Quadratic terms	3	77.011	25.670	91.379	0.011*
Linear terms	3	100.506	33.502	119.257	
Residual	2	0.562	0.281		
Total	8	178.079			

Table 14   ANOVA for Exercise 6.11(i)

(ii) If $\beta_1 = 2\beta_2$ it follows that $\beta_2 = \beta_1/2$, and the constrained model can be fitted by using $x_3 = x_1 + x_2/2$ in a simple linear regression model for y. A full Analysis of Variance in given in Table 15. We retain the null hypothesis of $\beta_1 = 2\beta_2$ with $p = 0.416$.

Source	df	SS	MS	F	p
Unconstrained	1	0.055	0.055	1.035	0.416
$\beta_1 = \beta_2$	1	22.685	22.685	424.009	
Residual	2	0.107	0.054		
Total	4	25.172			

Table 15   ANOVA for Exercise 6.11 (ii)

Chapter 7

7.1

(i) and (ii) produce exactly the same results (see Table 16).

Model	F-test	Action
X_1, X_2, X_3, X_4	$p = 0.289$	Remove X_3
X_1, X_2, X_3	$p = 0.007$	Retain X_4 and all other terms

Table 16   Stepwise and backward selection for Exercise 7.1

(iii) The results for forward selection are shown in Table 17. Note that the three methods do not produce the same results. Note also that whilst the results of these automated algorithms are interesting, they are probably insufficient in themselves to determine final model choice. Here, one might suggest X_1, X_2 offers a better choice of model.

Model	F-test	Action
Constant	$p = 0.000$	Add X_4
X_4	$p = 0.000$	Add X_1
X_4, X_1	$p = 0.06$	Do not add X_3

Table 17 Forward selection results for Exercise 7.1

7.2

(i) The plot of residuals versus fitted values is rather difficult to interpret (there are not really enough data). Really, it is only pragmatic experience which motivates this choice of square-root transformation.

(ii–iii) The Analysis of Variance table is shown in Table 18. We have conclusive evidence for differences between the treatments, but blocking does not appear to be particularly effective at reducing the experimental error. These conclusions are largely reinforced once we repeat the analysis using the square-root transformation (Table 19), although there is some suggestion that blocking is now more effective in this case.

Source	df	SS	MS	F	p
Treatment	4	597514	149379	45.898	0.000
Block	3	10244	3415	1.049	0.406
Residual	12	39055	3255		
Total	19	646813			

Table 18 ANOVA for Exercise 7.2

Source	df	SS	MS	F	p
Treatment	4	866.96	216.74	53.300	0.000
Block	3	22.61	7.54	1.854	0.191
Residual	12	4.07			
Total	19	938.365			

Table 19 ANOVA for Exercise 7.2 after square-root transformation

7.3

(i) A plot of fitted values against residuals gives some suggestion of 'funnelling out' of residuals and hence that a transformation might be appropriate.

(ii-iii) Using the raw data, the constructed ANOVA table is shown in Table 20, and using the transformation, in Table 21. In both cases we return broadly the same conclusions. We have strong evidence for differences between the Treatments, with Treatments C and D seeming to lead to fewer spoiled plants. In both cases the term for blocks is non-significant, although in both cases $F \approx 2$, suggesting that blocking is at least partially successful in reducing residual variation.

Source	df	SS	MS	F	p
Treatment	3	3116.56	1038.85	17.204	0.000
Block	5	626.66	125.33	2.076	0.126
Residual	15	905.77	60.38		
Total	23	4648.99			

Table 20   ANOVA for Exercise 7.3

Source	df	SS	MS	F	p
Treatment	3	0.445	0.148	13.378	0.000
Block	5	0.109	0.022	1.969	0.142
Residual	15	0.166	0.011		
Total	23	0.721			

Table 21   ANOVA for Exercise 7.3 using transformed data.

7.4

(i) Residual plot shows clear 'funnelling out' of residuals.

(ii) The means (and ranges in brackets) for each type are Type I 670.75 (663), Type II 1701.25 (1430), Type III 30775 (24400), Type IV 9395.883 (9440). On taking logs, these numbers become Type I 6.453 (0.998), Type II 7.417 (0.900), Type III 10.312 (0.829), Type IV 9.123 (0.939). These findings are interpreted in Snedecor and Cochran (1989) to mean that the log-transformation is successful in making, at least approximately, the ranges of the transformed data equal and uncorrelated with the means.

(iii) The ANOVA tables obtained are shown in Tables 22–23. We have significant evidence for differences in the numbers of the four different types of plankton. It appears that in order of increasing numbers we have Type III, Type IV, Type II, Type I. The different hauls also seem to account for a sub-

stantial amount of the variation observed, particularly when we consider the
logged data.

Source	df	SS	MS	F	p
Type	3	7035039590	2345013197	228.708	0.000
Haul	11	215279055	19570823	1.909	0.075
Residual	33	338358557	10253290		
Total	47	7588677202			

Table 22   ANOVA for Exercise 7.4

Source	df	SS	MS	F	p
Type	3	106.938	35.646	965.745	0.000
Haul	11	1.789	0.163	4.407	0.000
Residual	33	1.218	0.037		
Total	47	109.945			

Table 23   ANOVA for Exercise 7.4 using logged data

7.5

Both transformations yield ANOVA tables similar to Table 23 in that both
type and haul are seen to be highly significant (see Tables 24–25). However,
in the case of the square-root transformation, there remains the suggestion of
funnelling out of residuals.

Taylor's transformation. The mean (and variance) by type are Type I 670.75
(54719.46), Type II 1701.25 (127118.77), Type III 30775 (447711662.68), Type
IV 9395.833 (5410443.47). This leads to the estimate $\gamma = 1.832$, which is close
to the logarithmic transformation recovered in the case $\gamma = 2$. The Box–Cox
transformation gives $\lambda = 0.084$. Residual plots for both Taylor's and the logarithmic transformation appear reasonable.

Source	df	SS	MS	F	p
Type	3	163159	54386	617.821	0.000
Haul	11	3052	277	4.407	0.005
Residual	33	2905	88		
Total	47	169116.3			

Table 24   ANOVA for Exercise 7.5. Square-root transformation

Source	df	SS	MS	F	p
Type	3	441.16	147.05	1062.616	0.000
Haul	11	7.09	0.64	4.656	0.000
Residual	33	4.57	0.14		
Total	47	452.818			

Table 25 ANOVA for Exercise 7.5: Taylor's power law

7.6
(i) Since it is assumed $V - f(\mu) = F(\mu)(U - \mu)$, it follows that

$$
\begin{aligned}
E(\dot{V} - f(\mu))(V - f(\mu))^T &= E(F(\mu)(U - \mu))(F(\mu)(U - \mu))^T, \\
&= F(\mu)E\left((U - \mu)(U - \mu)^T\right)F(\mu)^T, \\
&= F(\mu)\Sigma_U F(\mu)^T.
\end{aligned}
$$

(ii) $E(V) \approx \sqrt{\mu}$, $\operatorname{var}(V) \approx 1$.
(iii) $E(V) \approx \log(\mu + 1)$, $\operatorname{var}(V) \approx \frac{\mu}{(1+\mu)^2}$, $\operatorname{var}(V) \approx \frac{1}{\mu}$ for large μ.

7.7
Transform to a linear model using $\log(y) = \log(\alpha) + \beta \log(x)$. The resulting linear model gives parameter estimates and estimated standard errors for $a = \log(\alpha)$ and β. We wish to estimate $\alpha = e^a$. We have that $\hat{\alpha} = \exp\{\hat{a}\}$ with e.s.e $\exp\{\hat{a}\}(\text{e.s.e}(a))$.

7.9
See Table 26.

7.10
Repeating the analysis of Exercise 3.4 leaves us with the model $Y = a + bZ$. Plotting residuals against fitted values suggests funnelling out of residuals, with some suggestion that the standard deviation is proportional to the mean (a plot of absolute residuals against fitted values can be helpful here). This suggests a logarithmic transformation of the response to stabilise variances. Analysis using t-statistics again suggests the simple model $\log(Y) = a + bZ$.

A plot of residuals versus fitted values suggests horizontal banding of residuals apart from three apparently outlying observations (observations 2, 43, 61). A plot of standardised residuals shows reasonable normality. A plot of residuals against leverage shows that of these outlying observations only observation 2 appears to have a relatively large leverage. All observations have Cook's distances within the control limits in the R plot. There are no short cuts, and if we want to be really thorough we should examine the effect of deleting each of the

Residual	Int. Stud. Res.	Ext. Stud Res.	Leverage	Cook's Dist.
3.653	1.315	1.361	0.219	2.216
3.500	1.239	1.270	0.192	1.901
2.596	0.906	0.899	0.168	0.986
-3.341	-1.139	-1.154	0.128	1.487
-4.368	-1.464	-1.547	0.099	2.378
-1.542	-0.514	-0.497	0.088	0.289
-1.695	-0.562	-0.546	0.080	0.344
-2.159	-0.714	-0.699	0.074	0.551
-0.400	-0.134	-0.129	0.099	0.020
-2.558	-0.872	-0.862	0.128	0.871
-1.661	-0.572	-0.556	0.147	0.384
5.115	1.785	1.994	0.168	3.829
3.202	1.134	1.149	0.192	1.591
-0.342	-0.123	-0.118	0.219	0.019

Table 26   Results for Exercise 7.9

three putative outliers in turn. It appears that deleting observations does not have a particularly large effect (the regions constructed by taking two e.s.e's either side of each estimate all intersect – see Table 27). In summary, it appears that the model $\log(Y) = a + bZ$ and estimated using all given observations is the most sensible approach.

Deleted observations	Parameter estimates (e.s.e)
None	$a = 1.613$ (0.232), $b = 0.530$ (0.073)
2	$a = 1.504$ (0.242), $b = 0.569$ (0.077)
43	$a = 1.583$ (0.139), $b = 0.555$ (0.043)
61	$a = 1.637$ (0.223), $b = 0.528$ (0.070)
2, 43	$a = 1.456$ (0.140), $b = 0.600$ (0.045)
2, 61	$a = 1.523$ (0.233), $b = 0.568$ (0.074)
43, 61	$a = 1.607$ (0.121), $b = 0.553$ (0.038)
2, 43, 61	$a = 1.478$ (0.120), $b = 0.599$ (0.038)

Table 27   Effects of deleting observations in Exercise 7.10

Chapter 8

8.1
Normal

$$
\begin{aligned}
f(y;\theta,\phi) &= \frac{1}{\sqrt{2\pi\phi}}\exp\left(\frac{-(y-\theta)^2}{2\phi^2}\right), \\
&= \exp\left(\frac{-y^2+2\theta y+\theta^2}{2\phi^2}-\frac{1}{2}\log(2\pi\phi)\right), \\
&= \exp\left(\frac{y\theta-\theta^2/2}{\phi}-\frac{1}{2}\left(y^2+\log(2\pi\phi)\right)\right).
\end{aligned}
$$

Poisson
If $Y\sim\text{Po}(\lambda)$ then $f(y;\theta,\phi)=\exp\left(-\lambda+y\log(\lambda)-\log(y!)\right)$.
Binomial
If $ny\sim\text{Bi}(n,p)$ then

$$
\begin{aligned}
f(y;\theta,\phi) &= \exp\left(\log\binom{n}{ny}+ny\log(p)+(n-ny)\log(1-p)\right) \\
&= \exp\left(ny\log\left(\frac{p}{1-p}\right)+n\log(1-p)+\log\binom{n}{ny}\right).
\end{aligned}
$$

Setting $\phi/\omega=n^{-1}$ it follows that $\log(p/(1-p))=\theta$ giving $p=e^\theta/(1+e^\theta)$.
Further, we see that $-b(\theta)=\log(1-p)$ giving $b(\theta)=\log(1+e^\theta)$.
Gamma
From the form of a $\Gamma(\alpha,\beta)$ density we have that

$$
f(y;\theta,\phi)=\exp\left(-\beta y+(\alpha-1)\log(y)+\alpha\log\beta-\log\Gamma(\alpha)\right).
$$

Comparing coefficients, we see that $-\beta=\theta/\phi$ to give

$$
\exp\left(\frac{\theta y}{\phi}+(\alpha-1)\log y+\alpha\log\left(\frac{-\theta}{\phi}\right)-\log\Gamma(\alpha)\right).
$$

Set $\phi=1/\alpha$ to give

$$
\exp\left(\frac{\theta y+\log(-\theta)}{\phi}+\left(\frac{1}{\phi}-1\right)\log y-\frac{\log\phi}{\phi}-\log\Gamma\left(\frac{1}{\phi}\right)\right).
$$

The stated result follows, with $b(\theta)=-\log(-\theta)$.

8.2
Poisson
Written as a function of the μ_i the log-likelihood function is

$$
l(\mu|\phi,y)=\sum_i-\mu_i+y_i\log(\mu_i)+C.
$$

$$D(\hat{\mu}|\phi, y) = 2\left[\sum_i -y_i + y_i \log(y_i) - \left(\sum_i -\hat{\mu}_i + y_i \log(\hat{\mu}_i)\right)\right]$$
$$= 2\left[\sum_i y_i \log\left(\frac{y_i}{\hat{\mu}_i}\right) - (y_i - \hat{\mu}_i)\right].$$

Binomial

We have that $n_i y_i \sim \mathrm{Bi}(n_i, \mu_i)$ and written as a function of the μ_i the log-likelihood function is

$$l(\mu|\phi, y) = \sum_i n_i y_i \log(\mu_i) + n_i(1 - y_i) \log(1 - \mu_i) + C.$$

We can write $D(\hat{\mu}|\phi, y)$ as

$$2\sum_i n_i \left[y_i \log(y_i) + (1 - y_i) \log(1 - y_i) - y_i \log(\hat{\mu}_i) - (1 - y_i) \log(1 - \hat{\mu}_i)\right],$$

which simplifies to give

$$2\sum_i n_i \left\{y_i \log\left(\frac{y_i}{\hat{\mu}_i}\right) + (1 - y_i) \log\left(\frac{1 - y_i}{1 - \hat{\mu}_i}\right)\right\}.$$

Gamma

The likelihood function can be written as

$$\prod_i \frac{\left(\frac{\alpha}{\mu_i}\right)^\alpha y_i^{\alpha - 1} e^{-(\alpha/\mu_i) y_i}}{\Gamma(\alpha)}.$$

The log-likelihood function becomes

$$l(\mu|\phi, y) = \sum_i -\alpha \log(\mu_i) - \left(\frac{\alpha}{\mu_i}\right) y_i + C.$$

$$D(\hat{\mu}|\phi, y) = \frac{2}{\alpha} \sum_i \left[-\alpha \log(y_i) - \left(\frac{\alpha}{y_i}\right) y_i + \alpha \log(\hat{\mu}_i) + \left(\frac{\alpha}{\hat{\mu}_i}\right) y_i\right],$$
$$= 2\sum_i \log\left(\frac{\hat{\mu}_i}{y_i}\right) + \frac{y_i - \hat{\mu}_i}{\hat{\mu}_i}.$$

8.3

The Poisson log-linear model gives a residual deviance of 3606.4 on 33 df with p-value of 0.000. The χ^2 test statistic is 3712.452 also on 33 df Results are similar in both cases and give overwhelming evidence of an association between haul and number caught, in common with the results of Exercise 7.4.

8.4

(i) Residual deviance = 1.923. Null deviance = 163.745. χ^2 goodness–of–fit test gives $p = 0.750$. Test for significance of log(dose) gives $\chi^2 = 161.821$, $p = 0.000$.

(ii) Residual deviance = 5.261. Null deviance = 163.744. χ^2 goodness–of–fit test gives $p = 0.262$. Test for significance of log(dose) gives $\chi^2 = 158.484$, $p = 0.000$.

(iii) Transform to a linear model using the arc-sine transformation (Exercise 7.3). The resulting fit appears reasonable and the t-test is significant ($p = 0.001$). In summary, all three models give conclusive evidence that mortality is increasing with dose.

8.5

The residual deviance is 3468.5, $\hat{\phi} = 204.03$ and the residual deviance of the null model is 3523.3. The F-statistic is $(3523.3 - 3468.5)/(612.09) = 0.090$ and comparing to $F_{3,16}$ gives a p-value of 0.965. Similar to Exercise 7.2, we conclude that blocking does not seem to be particularly effective at reducing the experimental error.

8.6

The model with blocks and treatments gives a residual deviance of 20.458 on 15 df which is not really large enough to suggest over-dispersion. The χ^2 test of goodness-of-fit gives a p-value of 0.155. The model with blocks only gives a residual deviance of 81.250 on 18 df The χ^2 test then gives $\chi^2 = 60.792$ on 3 df and a p-value of 0.000. The null model has a residual deviance of 94.723 on 23 df The χ^2 test then gives 13.473 on 5 df and a p-value of 0.019, suggesting that blocking is effective in reducing experimental error. Using corner-point constraints, the parameter estimates obtained are shown in Table 28. Since these are greater than 3 in absolute size, these are clearly significant when compared to normal or t-distributions. The suggestion is that Treatments B-D give lower probabilities of unusable corn than Treatment A, with Treatments C and D best. Conclusions from Exercise 7.3 are broadly similar.

Parameter	Estimate	e.s.e	t/z
Treatment B	-0.690	0.210	-3.279
Treatment C	-1.475	0.243	-6.078
Treatment D	-1.556	0.248	-6.286

Table 28 Parameter estimates and e.s.e for Exercise 8.6

8.7

The full model is age*club, with a residual deviance of 0.005 on 10 df with $\hat{\phi} = 5.03 \times 10^{-4}$. The null model has residual deviance of 0.017 on 13 df The F-test gives $F = 7.754$, and a p-value of 0.006 when compared to $F_{3,10}$. Thus the full model offers an improvement over the null model. Next we test for the significance of the interaction term. The age+club model gives a residual deviance of 0.005 on 11 df with $\hat{\phi} = 4.73 \times 10^{-4}$. The F-test gives $F = 0.340$, and a p-value of 0.573 when compared to $F_{1,10}$, so there is no evidence for an interaction between age and club status. Next we test for the need for two intercepts. The model with age alone gives a residual deviance of 0.015 on 12 df The F-test gives $F = 19.780$ and a p-value of 0.001 against $F_{1,11}$. Thus the chosen model is linear in time but with different intercepts according to club status. The fitted equation, using the canonical reciprocal link, is

$$\frac{1}{\text{time}} = 1.331 \times 10^{-2} - 5.597 \times 10^{-5}(\text{age}) + 9.643 \times 10^{-4}(\text{club status}).$$

8.8

The density function can be written as

$$\exp\left(\frac{-\lambda y}{2\mu^2} + \frac{\lambda}{\mu} - \frac{\lambda}{2y} + \frac{1}{2}\log\left(\frac{\lambda}{2\pi y^3}\right)\right).$$

This gives $\theta/\phi = -\lambda\mu^{-2}$ and $-b(\theta)/\phi = \lambda/\mu$. Set $\phi = \lambda^{-1}$. It follows that $\theta = -\mu^{-2}$, and since $-b(\theta) = 1/\mu$ that $b(\theta) = -\sqrt{-2\theta}$. Finally we can read off that

$$c(y,\phi) = -\frac{1}{2}\left(\log\left(\frac{\lambda}{2\pi y^3}\right) + \frac{\phi}{y}\right).$$

8.9

The full model is age*club with a residual deviance of 5.445×10^{-5} on 10 df with $\hat{\phi} = 5.498 \times 10^{-6}$. The null model has residual deviance of 1.854×10^{-4} on 12 df The F-test gives $F = 7.937$, and a p-value of 0.005 when compared to $F_{3,10}$. Thus the full model offers an improvement over the null model. Next we test for the significance of the interaction term. The age+club model gives a residual deviance of 5.637×10^{-5} on 11 df with $\hat{\phi} = 5.175 \times 10^{-6}$. The F-test gives $F = 0.350$, and a p-value of 0.567 when compared to $F_{1,10}$, so there is no evidence for an interaction between age and club status. Next we test for the need for two intercepts. The model with only age gives a residual deviance of 1.611×10^{-5} on 12 df The F-test gives $F = 20.251$ and a p-value of 0.001 against $F_{1,11}$. Thus the chosen model is linear in time but with different intercepts according to club status. The fitted equation, using the canonical μ^{-2} link is

$$\frac{1}{\text{time}^2} = 1.727 \times 10^{-4} - 1.254 \times 10^{-6}(\text{age}) + 2.163 \times 10^{-5}(\text{club status}).$$

8.10

Gamma model. We set $1/t = (a + bx + c)$, where c corresponds to a dummy variable indicating club membership. We have that $t = (a + bx + c)^{-1}$. Differentiating we obtain

$$\frac{dt}{dx} = \frac{-b}{(a + bx + c)^2}.$$

Plugging in the values of a, b, and c suggested by Exercise 8.7 gives

$$\frac{dt}{dx}(63) = 0.48 \quad \text{when} \quad c = 1,$$

and

$$\frac{dt}{dx}(63) = 0.58 \quad \text{when} \quad c = 0,$$

This implies that the first author is losing around 29 seconds a year on the half-marathon through age alone, very close to the Rule of Thumb.

Inverse Gaussian model. Set $1/t^2 = (a + bx + c)$ so that $t = (a + bx + c)^{-1/2}$. Differentiating gives

$$\frac{dt}{dx} = -\frac{b}{2}(a + bx + c)^{-\frac{3}{2}}.$$

Plugging in the values of a, b, and c suggested by Exercise 8.9 gives

$$\frac{dt}{dx}(63) = 0.506 \quad \text{when} \quad c = 1,$$

and

$$\frac{dt}{dx}(63) = 0.691 \quad \text{when} \quad c = 0,$$

This suggests losing around 30 seconds a year through age alone, very close to the above.

Dramatis Personae: Who did what when

H. Akaike (1927–), Akaike Information Criterion (AIC), 1974 [§5.2.1]. (Hirotugu Akaike was awarded the Kyoto Prize in Mathematical Sciences in 2006.)
Thomas Bayes (1702–1761), Bayes's Theorem, 1764 [§9.1].
David Blackwell (1919–2008), Rao–Blackwell Theorem, 1947 [§4.5].
Raj Chandra Bose (1901–1987), estimability, 1944 [§3.3].
George E. Box (1919–), Box–Cox Transformation, 1964 [§7.2]. Box-Jenkins models of time series, 1970 [§9.4]. (George Box married Fisher's daughter Joan Fisher Box, author of Fisher's biography *R. A. Fisher, The life of a scientist*, Wiley, 1978.)
Edgar Buckingham (1867–1940), Buckingham's Pi Theorem, 1914 [§8.2].
William G. Cochran (1909–1980), Cochran's Theorem, 1934 [§3.5].
D. R. (Sir David) Cox (1924–), Box–Cox transformation, 1964 [§7.2], proportional hazards, 1972 [§9.5], orthogonal parameters, 1987 [§5.1.1].
Harald Cramér (1893–1985), Cramér–Wold device, 1936 [§4.3].
Albert Einstein (1879–1955), Principle of Parsimony [§4.1.1].
Francis Ysidro Edgeworth (1845–1926), Edgeworth's Theorem, 1893 [§4.4].
Robert F. Engle (1942–), cointegration, 1991 [§9.4].
Gustav Fechner (1801–1887), Fechner's Law (power law), 1860 [§8.2].
R. A. (Sir Ronald) Fisher (1890–1962), likelihood, 1912 [§1.6], density of r^2, 1915 [§7.3], F-distribution, 1918 [§2.6], ANOVA, 1918 [Ch. 2], sufficiency, 1920 [§4.4.1], z-transformation, 1921 [§7.3], method of maximum likelihood, 1922 [§1.6], Fisher's Lemma, 1925 [§2.5], ANCOVA, 1932 [§5.2], ancillarity, 1934 [§5.1], information matrix, 1934 [§3.3], design of experiments, 1935 [§9.3], method of scoring, 1946 [§8.1].
Sir Francis Galton (1822–1911), *Hereditary genius*, 1869 [§1.3].
Carl Friedrich Gauss (1777–1855), least squares, 1795 [§1.2], Gauss–Markov Theorem, 1823 [§3.3].

Roy C. Geary (1896–1983), Geary's Theorem, 1936 [§2.5].

Irving John Good (1916–2009), penalised likelihood, roughness penalty, 1971 [§9.1], Bayesian statistics, 1950 on [§9.1]. (Jack Good was Alan Turing's statistical assistant working on deciphering German radio traffic at Bletchley Park in World War II, and one of the fathers of both Bayesian statistics and artificial intelligence.)

C. W. J. (Sir Clive) Granger (1934–2009), cointegration, 1991 [§9.4].

Friedrich Robert Helmert (1843–1917), Helmert's transformation, 1876 [§2.4].

Charles Roy Henderson (1911–1989), mixed models, 1950 on [§9.1].

Carl Gustav Jacob Jacobi (1804–1851), Jacobian, 1841 [§2.2].

S. Kołodziejczyk (-1939), Kołodziejczyk's Theorem: F-test for linear hypotheses, 1935 [§6.2]. (Stanisław Kołodziejczyk (pronounced 'Kowod*jay*chick'), a pupil of Jerzy Neyman, was killed fighting in the Polish Army against the German invasion at the beginning of World War II.)

Daniel Gerhardus Krige (1919–), kriging (spatial regression), 1951 [§9.2.1].

Joseph Louis Lagrange (1736–1813), Lagrange multipliers, 1797 [§6.1].

Adrien-Marie Legendre (1752–1833), least squares, 1805 [§1.2].

Andrei Andreyevich Markov (1856-1922), Gauss–Markov Theorem, 1912 [§3.3].

Georges Matheron (1930–2000), kriging, 1960s [§9.2.1].

John Ashworth Nelder (1924–2010), GenStat®, 1966 on, Generalised Linear Models, 1972, GLIM®, 1974 on, Hierarchical GLMs, 1996 [Ch. 8].

Sir Isaac Newton (1642–1727), *Principia*, 1687 [§1.2].

William of Ockham (d. c1349), Occam's Razor [§4.4.1].

Georg Simon Ohm (1787–1854), Ohm's Law, 1826 [§1.1].

Karl Pearson (1857–1936), Chi-squared test, 1900 [§8.4].

Pythagoras of Croton (d. c497 BC), Pythagoras's Theorem [§3.6].

Calyampudi Radhakrishna Rao (1920–), Rao–Blackwell Theorem, 1945 [§4.5].

Nancy Reid (1952–), orthogonal parameters, 1987 [§5.1].

Isaac J. Schoenberg (1903–1990), splines, 1946 on [§9.2].

Issai Schur (1875–1941), Schur complement, 1905 [§4.5, §9.1].

Shayle Robert Searle (1928–), mixed models, 1950s on [§9.1].

John Wilder Tukey (1915–2000), exploratory data analysis (EDA), 1977 [§1.1, §2.6].

Grace Wahba (1934–), Splines in statistics, 1970s [§9.2].

Sir Gilbert Walker (1868–1958), Yule–Walker equations, 1931 [§9.4].

Hermann Wold (1908–1992), Cramér–Wold device, 1936 [§4.3].

Frank Yates (1902–1994), design of experiments, 1930s on [§9.3].

George Udny Yule (1871–1951), Yule–Walker equations, 1926 [§9.4] spurious regression, 1927 [§9.4].

Bibliography

[1] Aitkin, Murray, Anderson, Dorothy, Francis, Brian and Hinde, John (1989) *Statistical modelling in GLIM*, Oxford University Press [182].

[2] Atkinson, Anthony C. (1985) *Plots, transformations and regression: An introduction to graphical methods of diagnostic regression analysis*, Oxford University Press [166, 177].

[3] Atkinson, A. C. and Donev, A. N. (1992) *Optimum experimental designs*, Oxford University Press, Oxford [216].

[4] Atkinson, Anthony C. and Riani, Marco (2000) *Robust diagnostic regression analysis*, Springer [177].

[5] Bingham, N. H. and Rashid, S. N. (2008) *The effect of ageing on athletic performance.* Preprint [93, 188, 204].

[6] Bishop, Y. M. M., Fienberg, S. E., and Holland, P. W. (1995) *Discrete multivariate analysis: Theory and practice*, MIT Press, Cambridge, MA [195].

[7] Blyth, T. S. and Robertson, E. F. (2002a) *Basic linear algebra*, Springer, SUMS [vi, 36, 64, 80, 82, 114].

[8] Blyth, T. S. and Robertson, E. F. (2002b) *Further linear algebra*, Springer, SUMS [vi, 39, 64, 69, 75, 80, 109].

[9] Box, G. E. P. and Cox, D. R. (1964) An analysis of transformations, *J. Royal Statistical Society B* **26**, 211–246 [169].

[10] Box, G. E. P. and Jenkins, G. W. (1970) *Time series analysis, forecasting and control*, Holden–Day (2nd ed. 1994 with Reinsel, G. C.)[220].

[11] Brockwell, P. J. and Davis, R. A. (2002) *Introduction to time series and forecasting*, 2nd ed., Springer [220].

[12] Brown, Lawrence D. (1986) *Fundamentals of statistical exponential families, with applications in statistical decision theory*, IMS Lecture Notes – Monograph Series 9, Institute of Mathematical Statistics, Hayward CA [183].

[13] Carroll, R. J. and Ruppert, D. (1988) *Transformation and weighting in regression*, Chapman and Hall [125].

[14] Casella, G. and Berger, R. L. (1990) *Statistical inference*, Duxbury Press [113].

[15] Collett, David (2003) *Modelling binary data*, 2nd ed., Chapman and Hall/CRC [191].

[16] Cook, R. Dennis and Weisberg, Sanford (1982) *Residuals and influence in regression*, Chapman and Hall [167].

[17] Copson, E. T. (1935), *Functions of a complex variable*, Oxford University Press [21, 37].

[18] Cox, D. R. and Oakes, D. (1984) *Analysis of survival data*, Chapman and Hall/CRC [224].

[19] Cox, D. R. and Reid, N. (1987) Parameter orthogonality and approximate conditional inference (with discussion). *J. Roy. Statist. Soc. B* **49**, 1–39.

(Reprinted in *Selected statistical papers of Sir David Cox*, ed. D. J. Hand and A. M. Herzberg, Volume 2, 309–327, Cambridge University Press, (2005)) [134].

[20] Cox, D. R. and Snell, E. J. (1989) *Analysis of binary data*, 2nd ed., Chapman and Hall/CRC [191].

[21] Cramér, Harald (1946) *Mathematical methods of statistics*, Princeton University Press [71, 194, 195].

[22] Crawley, Michael J. (1993) *GLIM for Ecologists*, Blackwell, Oxford [182].

[23] Crawley, Michael J. (2002) *Statistical computing: An introduction to data analysis using S-Plus*, Wiley [48, 163, 182].

[24] Cressie, N. A. C. (1993) *Statistics for spatial data*, Wiley [215].

[25] de Boor, C. (1978) *A practical guide to splines*, Springer [212].

[26] Diggle, P. J., Heagerty, P., Lee, K. Y., and Zeger, S. L. (2002) *Analysis of longitudinal data*, 2nd ed., Oxford University Press (1st ed., Diggle, Liang and Zeger, 1994) [203].

[27] Dineen, Seán (2001) *Multivariate calculus and geometry* 2nd ed., Springer, SUMS [36, 150].

[28] Dobson, Annette, J. and Barnett, Adrian, G. (2003) *An introduction to Generalised Linear Models*, 3rd ed., Chapman and Hall/CRC [196].

[29] Draper, Norman R. and Smith, Harry (1998) *Applied regression analysis*, 3rd ed., Wiley [28, 77, 105, 125, 164, 168].

[30] Fienberg, Stephen E. (1980) *The analysis of cross-classified categorical data*, 2nd ed., MIT Press, Cambridge MA [195].

[31] Fisher, R. A. (1958) *Statistical methods for research workers*, 13th ed., Oliver and Boyd, Edinburgh [130].

[32] Focken, C. M. (1953) *Dimensional methods and their applications*, Edward Arnold, London [171].

[33] Garling, D. J. H. (2007) *Inequalities: a journey into linear analysis*, Cambridge University Press [18].

[34] Golub, Gene H. and Van Loan, Charles F. (1996) *Matrix computations*, 3rd ed., Johns Hopkins University Press, Baltimore MD [65, 69].

[35] Green, P. J. and Silverman, B. W. (1994) *Non–parametric regression and Generalised Linear Models: A roughness penalty approach*, Chapman and Hall [213].

[36] Grimmett, Geoffrey and Stirzaker, David (2001) *Probability and random processes*, 3rd ed., Oxford University Press [8, 21, 194, 196].

[37] Haigh, John (2002) *Probability models*, Springer, SUMS [vi, 8, 16, 18, 19, 20, 34, 55, 115, 117, 196, 208].

[38] Halmos, P. R. (1979) *Finite–dimensional vector spaces*. Undergraduate Texts in Mathematics, Springer [75, 86, 109].

[39] Hand, D. J. (2004) *Measurement theory and practice: The world through quantification*, Edward Arnold [188].

[40] Healy, M. J. R. (1956) *Matrices for statistics*, Oxford University Press [127].

[41] Hirschfeld, J. W. P. (1998) *Projective geometries over finite fields*, Oxford University Press [218].

[42] Howie, John M. (2001) *Real analysis*, Springer, SUMS [18].

[43] Howie, John M. (2003) *Complex analysis*, Springer, SUMS [21].

[44] Huber, P. J. (1981) *Robust statistics*, Wiley, New York [156, 166].

[45] Jeffreys, Harold (1983) *Theory of probability*, 3rd ed., Oxford University Press [102].

[46] Joiner, Brian L. and Ryan, Barbara F. (2000) *MINITAB student handbook*, 4th ed., Duxbury Press, Boston MA [104].

[47] Kendall, Sir Maurice and Stuart, A. (1977) *The advanced theory of statistics, Volume 1: Distribution theory*, 4th ed., Charles Griffin, London [37, 39, 41, 85, 146, 172].

[48] Kendall, Sir Maurice and Stuart, A. (1979) *The advanced theory of statistics, Volume 2: Inference and relationship*, 4th ed., Charles Griffin, London [25, 182].

[49] Mardia, K. V., Kent, J. T. and Bibby, J. M. (1979) *Multivariate analysis*, Academic Press [109].

[50] Markov, A. A. (1912) *Wahrscheinlichkeitsrechnung*, Teubner, Leipzig [73].

[51] McConway, K. J., Jones, M. C. and Taylor, P. C. (1999) *Statistical modelling using GENSTAT*, Arnold, London and The Open University [182].

[52] McCullagh, P. and Nelder, J. A. (1989) *Generalised Linear Models*, 2nd ed., Chapman and Hall [168, 181, 182, 185, 191, 197].

[53] Montgomery, Douglas, C. (1991) *Design and analysis of experiments*, 3rd ed., Wiley [216].

[54] Mosteller, Frederick and Wallace, David L. (1984) *Applied Bayesian and classical inference: The case of the Federalist papers*, Springer (1st ed., *Inference and disputed authorship: The Federalist*, Addison-Wesley, Reading MA, 1964) [198].

[55] Nelder, J. A., Lee, Y. and Pawitan, Y. (2006) *Generalised linear models with random effects: unified analysis via H-likelihood*, Chapman and Hall/CRC [189].

[56] Ostaszewski, Adam (1990) *Advanced mathematical methods*, Cambridge University Press [150].

[57] Plackett, R. L. (1960) *Regression analysis*, Oxford University Press [103].

[58] Plackett, R. L. (1974) *The analysis of categorical data*, Griffin, London [195].

[59] Rao, C. R. (1973) *Linear statistical inference and its applications*, 2nd ed., Wiley [71, 85, 120].

[60] Ruppert, David, Wand, M. P. and Carroll, R. J. (2003) *Semiparametric regression*, Cambridge University Press [211, 213, 214, 215].

[61] Searle, S. R. (1991) *Linear models*, Wiley [204].

[62] Searle, S. R. (1982) *Matrix algebra useful for statistics*, Wiley [204].

[63] Searle, S. R., Casella, G. and McCulloch, C. E. (1992) *Variance components*, Wiley, New York [55, 204].

[64] Seber, George A. F. and Lee, Alan, J. (2003) *Linear regression analysis*, 2nd ed., Wiley [159, 165].

[65] Snedecor, G. W. and Cochran, W. G. (1989) *Statistical methods*, 8th ed., Iowa State University Press, Iowa [48, 260].

[66] Stigler, Stephen M. (1986) *The history of statistics: The measurement of uncertainty before 1900*, Harvard University Press, Cambridge MA [22].

[67] Szegö, G. (1959) *Orthogonal polynomials*, AMS Colloquium Publications XXIII, American Mathematical Society [103].

[68] Taylor, L. R. (1961) Aggregation, variance and the mean, *Nature* **189**, 732–735 [173].

[69] Tukey, J. W. (1977) *Exploratory data analysis*, Addison–Wesley, Reading MA [42].

[70] Venables, W. N. and Ripley, B. D. (2002) *Modern applied statistics with S*, 4th ed., Springer [164, 170, 181, 182, 185, 199, 213, 214, 224].

[71] Wilkinson, J. H. (1965) *The Algebraic Eigenvalue problem*, Oxford University Press [100].

[72] Williams, David (2001) *Weighing the odds: A course in probability and statistics*, Cambridge University Press [115, 117, 123, 209].

[73] Woods, H., Steinour, H. H. and Starke, H. R. (1932): Effect of composition of Portland cement on heat evolved during hardening, *Industrial and Engineering Chemistry* **24**, 1207–1214 [175].

[74] Wynn, H. P. (1994) Jack Kiefer's contribution to experimental design *Annals of statistics*, **12**, 416–423 [218].

[75] Zhang, Fuzhen (2005) *The Schur complement and its applications*, Springer [120].

Index

Printed in Great Britain
by Amazon